# KINGS DETHRONED

A HISTORY OF THE EVOLUTION OF ASTRONOMY
FROM THE TIME OF THE ROMAN EMPIRE UP
TO THE PRESENT DAY; SHOWING IT TO BE AN
AMAZING SERIES OF BLUNDERS FOUNDED
UPON AN ERROR MADE IN THE
SECOND CENTURY B.C.

By
GERRARD HICKSON

Published by
THE HICKSONIA PUBLISHING CO.
2 Bride Court, Fleet Street, London, E.C.4.
and
21 Newberry Avenue, Stapleton, Staten Island,
New York, U.S.A.
1922

# KINGS DETHRONED

A history of the evolution of astronomy
from the time of the Roman Empire up
to the present day ; showing it to be an
amazing series of blunders founded upon
an error made in the second century B.C.

By
GERRARD HICKSON

# REFERENCE NOTES.

# PREFACE

In the year 1907 the author made a remarkable discovery which convinced him that the sun was very much nearer to the earth than was generally supposed. The fact he had discovered was demonstrated beyond all doubt, so that he was compelled to believe that—however improbable it might seem—astronomers had made a mistake when they estimated the distance of the sun to be ninety-three millions of miles.

He then proceeded to examine the means by which the sun's distance had been computed, and found an astounding error in the " Diurnal Method of Measurement by Parallax," which had been invented by Dr. Halley in the early part of the 18th century, and which was used by Sir David Gill in measuring the distance to the planet Mars in 1877 ; from which he deduced his solar parallax of 8.80".

Seeing that Sir Norman Lockyer had said that the distance to and the dimensions of everything in the firmament except the moon depends upon Sir David Gill's measurement to Mars, the author set himself the tremendous task of proving the error, tracing its consequences up to the present day, and also tracing it backwards to the source from which it sprang.

The result of that research is a most illuminating history of the evolution of astronomy from the time of the Roman Empire up to April 1922 ; which is now placed in the hands of the people in " Kings Dethroned."

The author has taken the unusual course of submitting these new and startling theories for the consideration of the general public because the responsible scientific societies in London, Washington and Paris, failed to deal with the detailed accounts of the work which he forwarded to them in the Spring of 1920. He believes that every newly-discovered truth belongs to the whole of *mankind, wherefore, if those whose business it is to consider his work fail in their duty he does not hesitate to bring it himself direct to the people, assured of their goodwill and fair judgment.*

*Astronomy has ever been regarded as a study only for the few, but now all its strange terms and theories have been explained in the most lucid manner in " Kings Dethroned," so that everyone who reads will acquire a comprehensive knowledge of the science.*

*The author takes this opportunity of assuring the reader that none esteems more highly than he, himself, the illustrious pioneers who devoted their genius to the building of astronomy, for he feels that even while pointing out their errors he is but carrying on their work, striving, labouring even as they did, for the same good cause of progress in the interests of all. On the other hand, he thinks that astronomers living at the present time might have used to better purpose the greater advantages which this century provides, and done all that he himself has done by fearless reasoning, devoted labour ; and earnest seeking after truth.*

*G. H.*

## Chapter One

## WHEN THE WORLD WAS YOUNG

THREE thousand years ago men believed the earth was supported on gigantic pillars. The sun rose in the east every morning, passed overhead, and sank in the west every evening ; then it was supposed to pass between the pillars under the earth during the night, to re-appear in the east again next morning.

This idea of the universe was upset by Pythagoras some five hundred years before the birth of Christ, when he began to teach that the earth was round like a ball, with the sun going round it daily from east to west ; and this theory was already about four hundred years old when Hipparchus, the great Greek scientist, took it up and developed it in the second century, B.C.

Hipparchus may be ranked among the score or so of the greatest scientists who have ever lived. He was the inventor of the system of measuring the distance to far off objects by triangulation, or trigonometry, which is used by our surveyors at the present day, and which is the basis of all the methods of measuring distance which are used in modern astronomy. Using this method of his own invention, he measured from point to point on the surface of the earth, and so laid the foundation of our present systems of geography, scientific map-making and navigation.

It would be well for those who are disposed to under-estimate the value of new ideas to consider how much the world owes to the genius of Hipparchus, and to try to conceive how we could have made progress—as we know it—without him.

TRIANGULATION.

The principles of triangulation are very simple, but because it will be necessary—as I proceed—to show how modern astronomers have departed from them, I will explain them in detail.

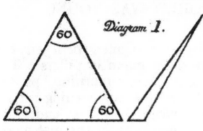

Diagram 1.

Every figure made up of three connected lines is a tri—or three-angle, quite regardless of the length of any of its sides. The triangle differs from all other shapes or figures in this :—that the value of its three angles, when added together, admits of absolutely no variation ; they always equal 180 degrees ; while —on the other hand—all other figures contain angles of 360 degrees or more. The triangle alone contains 180 degrees, and no other figure can be used for measuring distance. There is no alternative whatever, and therein lies its value.

It follows, then, that if we know the value of any two of the angles in a triangle we can readily find the value of the third, by simply adding together the two known angles and subtracting the result from 180. The value of the third angle is necessarily the remainder. Thus in our example (diagram 2) an angle of 90 degrees plus an angle of 60 equals 150, which shows that the angle at the distant object—or apex of the triangle— must be 30.

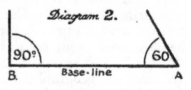

Diagram 2.

90°

B.

Base-line

60

A

Now if we know the length of the base-line A—B, in feet, yards, kilometres or miles, (to be ascertained by actual measurement), and also know the value of the two angles which indicate the direction of a distant object as seen from A. and B., we can readily complete the triangle and so find the length of its sides. In this way we can measure the height of a tree or church steeple from the ground level, or find the distance to a ship or lighthouse from the shore.

The reader will perceive that to obtain any measurement by triangulation it is absolutely necessary to have a base-line, and to know its length exactly. It is evident, also, that the length of the base-line must bear a reasonable proportion to the dimensions of the triangle intended; that is to say,—that the greater the distance of the object under observation the longer the base-line should be in order to secure an accurate measurement.

A little reflection will now enable the reader to realize the difficulties which confronted Hipparchus when he attempted to measure the distance to the stars.

It was before the Roman Conquest, when the geography of the earth was but little known, and there were none of the rapid means of travelling and communication which are at our disposal to-day. Moreover, it was in the very early days of astronomy, when there were few—if any—who could have helped Hipparchus in his work, while if he was to make a successful triangulation to any of the stars it was essential that he should have a base-line thousands of miles in length, with an observer at each end ; both taking observations to the same star at precisely the same second of time.

The times in which he lived did not provide the conveniences which were necessary for his undertaking, the conditions were altogether impossible, and so it is not at all surprising that he failed to get any triangulation to the stars.   As a result he came to the conclusion that they must be too far off to be measured, and said " the heavenly bodies are infinitely distant."

Such was the extraordinary conclusion arrived at by Hipparchus, and that statement of his lies at the root of astronomy, and has led its advocates into an amazing series of blunders from that day to this. The whole future of the science of astronomy was affected by Hipparchus when he said " the heavenly bodies are infinitely distant," and now, when I say

that it is not so, the fate of astronomy again hangs in
the balance.    It is a momentous issue which will be
decided in due course within these pages.

The next astronomer of special note is Sosigenes,
who designed the Julian Calendar in the reign of
Cæsar.   He saw no fault in the theories of Hipparchus,
but handed them on to Ptolemy, an Egyptian
astronomer of very exceptional ability, who lived in
the second century A.D.

Taking up the theories of his great Greek pre-
decessor after three hundred years, Ptolemy accepted
them without question as the work of a master ;   and
developed them.    Singularly gifted as he was to carry
on the work of Hipparchus, his genius was of a different
order, for while the Greek was the more original thinker
and inventor the Egyptian was the more accomplished artist in detail ;   and the more skilful in the
art of teaching.    Undoubtedly he was eminently fitted
to be the disciple of Hipparchus, and yet for that very
reason he was the less likely to suspect, or to discover,
any error in the master's work.

In the most literal sense he carried on that work,
built upon it, elaborated it, and established the
Ptolemaic System of astronomy so ably that it stood un-
challenged and undisputed for fourteen hundred years;
and during all those centuries the accepted theory of the
universe was that the earth was stationary, while the
sun, moon, stars and planets revolved around it daily.

Having accepted the theories of Hipparchus in the
bulk, it was but natural that Ptolemy should fail to
discover the error I have pointed out, though even
had it been otherwise it would have been as difficult
for him to make a triangulation to the stars in the
second century A.D., as it had been for the inventor
of triangulation himself three hundred years earlier.
However, it is a fact that he allowed the theory that
" the heavenly bodies are infinitely distant " to
remain unquestioned ;   and that was an error of
omission which was ultimately to bring about the
downfall of his own Ptolemaic system of astronomy.

## Chapter Two

## COPERNICUS AND GALILEO.

PTOLEMY's was still the astronomy of the world when Columbus discovered America, 1492, but there was living at that time—in the little town of Franenburg, in Prussia—a youth of 18, who was destined in later years to overthrow the astronomy of Hipparchus and Ptolemy, and to become himself the founder of a new theory which has since been universally accepted in its stead ; Nicholas Copernicus.

It is to be remembered that at that time the earth was believed to stand still, while the sun, moon, planets and stars moved round it daily from east to west, as stated by Ptolemy ; but this did not seem reasonable to Copernicus. He was a daring and original thinker, willing to challenge any theory—be it ever so long established—if it did not appear logical to him, and he contended that it was unreasonable to suppose that all the vast firmament of heavenly bodies revolved around this relatively little earth, but, on the contrary, it was more reasonable to believe that the earth itself rotated and revolved around an enormous sun, moving within a firmament of stars that were fixed in infinite space ; for in either case the appearance of the heavens would be the same to an observer on the surface of the earth.

This was the idea that inspired Nicholas Copernicus to labour for twenty-seven years developing the Heliocentric Theory of the universe, and in compiling the book that made him famous :—" De Revolutionibus Orbium Cœlestium," which was published in the last year of his life : 1543.

And now it is for us to very carefully study this fundamental idea of the Heliocentric theory, for there is an error in it.

Ptolemy had made it appear that the sun and stars revolved around a stationary earth, but Copernicus advanced the theory that it was the earth which revolved around a stationary sun, while the stars were fixed; and either of these entirely opposite theories gives an equally satisfactory explanation of the appearance of the sun by day and the stars by night. Copernicus did not produce any newly-discovered fact to prove that Ptolemy was wrong, neither did he offer any proof that he himself was right, but worked out his system to show that he could account for all the appearances of the heavens quite as well as the Egyptian had done, though working on an entirely different hypothesis; and offered his new Heliocentric Theory as an alternative.

He argued that it was more reasonable to conceive the earth to be revolving round the sun than it was to think of the sun revolving round the earth, because it was more reasonable that the smaller body should move round the greater. And that is good logic.

We see that Copernicus recognised the physical law that the lesser shall be governed by the greater, and that is the pivot upon which the whole of his astronomy turns; but it is perfectly clear that in building up his theories he assumed the earth to be much smaller than the sun, and also smaller than the stars; and that was pure assumption unsupported by any kind of fact. In the absence of any proof as to whether the earth or the sun was the greater of the two, and having only the evidence of the senses to guide him, it would have been more reasonable had he left astronomy as it was, seeing that the sun appeared to move round the earth, while he himself was unconscious of any movement.

When he supposed the stars to be motionless in space, far outside the solar system, he was assuming them to be infinitely distant; relying entirely upon the statement made by Hipparchus seventeen hundred years before. It is strange that he should have accepted this single statement on faith while he was in the very act of repudiating all the rest of the astronomy of Hipparchus and Ptolemy, but the fact remains that he did accept the " infinitely distant " doctrine without question, and that led him to suppose the heavenly bodies to be proportionately large ; hence the rest of his reasonings followed as a matter of course.

He saw that the Geocentric Theory of the universe did not harmonise with the idea that the stars were infinitely distant, and so far we agree with him. He had at that time the choice of two courses open to him :—he might have studied the conclusion which had been arrived at by Hipparchus, and found the error there ; but instead of doing that he chose to find fault with the whole theory of the universe, to overthrow it, and invent an entirely new astronomy to fit the error of Hipparchus !

It was a most unfortunate choice, but it is now made clear that the whole work of Copernicus depends upon the single question whether the ancient Greek was right or wrong when he said " the heavenly bodies are infinitely distant." It is a very insecure foundation for the whole of Copernican or modern astronomy to rest upon, but such indeed is the case.

. . . . . .

Some thirty years after the publication of the work of Copernicus, Tycho Brahe, the Danish astronomer, invented the first instrument used in modern astronomy. This was a huge quadrant nineteen feet in height (the forerunner of the sextant), which he used to very good purpose in charting out the positions of many of the more conspicuous stars. He differed with some of the details of the Prussian doctor's theory, but accepted it in the main ; and took no account whatever of the question of the distance of the stars.

Immediately following him came Johann Kepler,

and it is a very remarkable circumstance that this German philosopher, mystic and astrologer, should have been the founder of what is now known as Physical Astronomy. Believer as he was in the ancient doctrine that men's lives are pre-destined and mysteriously influenced by the stars and planets, he nevertheless sought to discover some physical law which governed the heavenly bodies. Having accepted the Copernican Theory that the sun was the centre of the universe, and that the earth and the planets revolved around it, it was but natural that all his reasonings and deductions should conform to those ideas, and so it is only to be expected that his conclusions dealing with the relative distances, movements and masses of the planets, which he laboured upon for many years, and which are now the famous " Laws of Kepler," should be in perfect accord with the Heliocentric Theory of Copernicus.

But, though the underlying principles of Kepler's work will always have great value, his conclusions cannot be held to justify Copernican astronomy, since they are a sequel to it, but—on the contrary— they will be involved in the downfall of the theory that gave them birth.

. . . . . . .

While the life work of Johann Kepler was drawing to a close, that of Galileo was just beginning, and his name is more widely known in connection with modern astronomy than is that of its real inventor, Nicholas Copernicus. Galileo adopted the Copernican theory with enthusiasm, and propagated it so vigorously that at one time he was in great danger of being burnt at the stake for heresy. In the year 1642 he invented the telescope, and so may be said to have founded the modern method of observing the heavens.

Zealous follower of Copernicus as he was, Galileo did much to make his theory widely known and commonly believed, and we may be sure that it was because he saw no error in it that other giants of astronomy who came after him accepted it the more readily. Nearly eighteen hundred years had passed since Hipparchus had said the heavenly bodies were infinitely distant, and still no one had questioned the accuracy of that statement, nor made any attempt whatever to measure their distance.

It is interesting to mention here an event which—at first sight—might seem unimportant, but which—now reviewed in its proper place in history—can be seen to have had a marked effect on the progress of astronomy as well as navigation. This was the publication of a little book called " The Seaman's Practice," by Richard Norwood, in the year 1637. At that time books of any kind were rare, and this was the first book ever written on the subject of measuring by triangulation. It was intended for the use of mariners, but there is no doubt that " The Seaman's Practice " helped King Charles II. to realise how the science of astronomy could be made to render valuable service to British seamen in their voyages of discovery, with the result that in 1675 he appointed John Flamsteed to make a special study of the stars, and to chart them after the manner of Tycho Brahe and Galileo, in order that navigators might guide their ships by the constellations over the trackless oceans.

That was how the British School of Astronomy came into existence, with John Flamsteed as the first Astronomer Royal, employing only one assistant, with whom he shared a magnificent salary of £70 a year ; and navigation owes much to the excellent work he did with an old-fashioned telescope, mounted in a little wooden shed on Greenwich Hill.

At about the same time the French School of Astronomy came into being, and the end of the seventeenth century began the most glorious period in the history of the science, when astronomers in England, France and Germany all contested strenuously for supremacy, and worshipped at the shrine of Copernicus.

## Chapter Three

## OLE ROEMER'S BLUNDER

AMONG the many ambitious spirits of that time, was one whose name is known only to a comparative few, nevertheless he has had a considerable influence both on astronomy and physics—Ole Roemer, best remembered for his observations of the Eclipses of Jupiter's Satellites.

A study of the records which have been made during more than 3,000 years shows that eclipses repeat themselves with clock-work regularity, so that a given number of years, months, days and minutes elapse between every two eclipses of a given kind ; but Ole Roemer observed that in the case of the eclipses of the satellites, or moons, of Jupiter, the period of time between them was not always the same, for they occurred 16½ minutes later on some occasions than on others. He therefore tried to account for this slight difference in time, and was led to some strange conclusions.

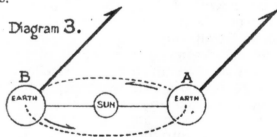

Diagram 3.

These eclipses occur at different seasons of the year, so that sometimes they can be seen when the earth is at A (see dia. 3), and at other times when the earth is at B, on the opposite side of the sun and the orbit, (according to Copernican Astronomy).

So Ole Roemer reflected that when the observer is at B, he is further from Jupiter than he is when the earth is at A, by a distance as great as the diameter of the orbit ; and that gave him a new idea, and a possible explanation.

He thought that although light appeared to be—for all ordinary purposes—instantaneous, it really must take an appreciable time to travel over the immense distance from Jupiter to the earth, just as a ship takes so long to travel a given distance at so many miles per hour. In that case the light from Jupiter's satellites would take less time to reach the observer when the earth is at A than it would require to reach him at B, on the further side of the orbit; and as a result of these reflections he reached the conclusion that the 16½ minutes difference in time was to be accounted for in that way.

Following up this idea, he decided that if it took 16½ minutes longer for light to travel the increased distance from one side of the orbit to the other, it would require only half the time to travel half that distance, so that it would travel as far as from the sun to the earth in 8¼ minutes. Therefore he gave it as his opinion that the distance to the sun was so tremendous that "Light"—travelling with almost lightning rapidity—took 8¼ minutes to cover the distance.

This ingenious hypothesis appealed strongly to the imagination of contemporary astronomers, so that they allowed it to pass without a sufficient examination, with the result that eventually it took its place among the many strange and ill-considered theories of astronomy. . . However, we ourselves will now do what should have been done in the days of Ole Roemer. We will stand beside him, as it were, and study these eclipses of Jupiter's satellites, just as he did, from the same viewpoint of Copernican astronomy ; and then we shall find whether his deductions were justified or not.

The eclipses are to be seen on one occasion when the observer (or earth) is at A, and on another occasion

when the observer (or earth) is at B, while the light of
Jupiter's satellites (or the image of the eclipse) is
supposed to cross the orbit at one observation but not
at the other.   It is important to note that the observer
at B will have to look in a direction toward the sun,
and across the orbit;  while the observer at A will
see the eclipse outward from the orbit;  in a direction
opposite to the sun. . .   Ole Roemer found that the
observer at B saw the eclipse $16\frac{1}{2}$ minutes later than
he would have seen it from A, and he believed that
this was because the image of the eclipse had a greater
distance to come to meet his eye.

Let us now consider diagram 4, which shows two
observers in the positions Ole Roemer supposed

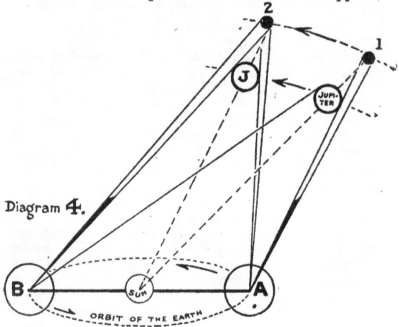

Diagram 4.

the earth to occupy at the respective observations.
We find that A would see the satellite in a state
of eclipse while it would be hidden from B by
the planet Jupiter; (triangle A, 1, B).  The planet and
its satellite are both moving round the sun toward

the east, as shown by the arrows, but the satellite is
like a moon, travelling round Jupiter; so that it
moves faster than the planet.    The satellite is eclipsed
by Jupiter only when the two are together on the same
line with the sun, (dotted lines), but, as time passes,
the satellite moves to the eastward of that line; it
passes Jupiter; and then it can be seen by the
observer at B. (triangle B, 2, A).

Thus it is that B sees the eclipse a few minutes later
than A, and that is the very simple explanation
which Ole Roemer overlooked.    It would be possible
to write a volume on this subject, and there are some
who would want to debate it at interminable length,
but in the end the explanation would prove to be just
this; which I prefer to leave in all its simplicity.
The $16\frac{1}{2}$ minutes difference in time is due to a difference
in the angles from which the eclipses are seen, and is
not in any way connected with distance; and so
the speculations of Ole Roemer concerning the
Velocity of Light and the probable distance to the
sun amount to nothing.

## Chapter Four

# GIANTS OF MODERN ASTRONOMY

BEFORE passing on to the more important part of this work, it is only just to record the fact that the first practical work in triangulation since the time of Hipparchus was performed by Jean Picard and J. and D. Cassini, between Paris and Dunkirk toward the end of the 17th century ; when Newton was working out his theories.

At this time the Copernican theory of astronomy was well established, and was accepted by all the scientific world, though it is probable that the public in general found it difficult to reconcile the idea of an earth careering through space at prodigious speed with common sense and reason. Even the most ardent followers of Copernicus and Galileo recognised this difficulty, and some strove to find a satisfactory explanation.

Nearly a hundred years ago Kepler had suggested that some kind of unknown force must hold the earth and the heavenly bodies in their places, and now Sir Isaac Newton, the greatest mathematician of his age, took up the idea and built the Law of Gravitation.

The name is derived from the Latin word " gravis," which means " heavy," " having weight," while the Law of Gravitation is defined as " That mutual action between masses of matter by virtue of which every such mass tends toward every other with a force varying directly as the product of the masses, and inversely as the square of their distances apart." Reduced to simplicity, gravitation is said to be " That which attracts every thing toward every other thing."

That does not tell us much ; and yet the little it does tell us is not true ; for a thoughtful observer knows very well that every thing is not attracted towards every other thing. . . The definition implies that it is a force ; but it does not say so, for that phrase "mutual action" is ambiguous, and not at all convincing.

The Encyclopædia Britannica tells us that "The Law of Gravitation is unique among the laws of nature, not only for its wide generality, taking the whole universe into its scope, but in the fact that, so far as is yet known, it is absolutely unmodified by any condition or cause whatever."

Here again we observe that the nature of gravitation is not really defined at all ; we are told that masses of matter tend toward each other, but no reason is given why they do so, or should do so ; while to say that "it is absolutely unmodified by any condition or cause whatever" is one of the most unscientific statements it is possible to make. There is not any thing or force in the universe that is absolute ! no thing that goes its own way and does what it will without regard to other forces or things. The thing is impossible; and it is not true; wherefore it has fallen to me to show where the inconsistency in it lies.

The name given to this mutual action means "weight," and weight is one of the attributes of all matter. Merely to say that anything is matter or material implies that it has weight, while to speak of weight implies matter. Matter and weight are inseparable, they are not laws, but elemental facts. They exist.

But it has been suggested that gravitation is a force, indeed we often hear it referred to as the force of gravitation ; but force is quite a different thing than weight, it is active energy expressed by certain conditions and combinations of matter. It acts.

All experience and observation goes to prove that material things fall to earth because they possess the attribute of weight, and that an object remains suspended in air or space only so long as its weight is overcome by a force, which is contrary. And when we realize these simple facts we see that gravitation is in reality conditioned and modified by every other active force, both great and small.

Again, gravitation is spoken of as a pull, an agent of attraction that robs weight of its meaning, something that brings all terrestrial things down to earth while at the same time it keeps the heavenly bodies in their places and prevents them falling toward each other or apart. The thing is altogether too wonderful, it is not natural; and the theory is scientifically unsound. . .

Every man, however great his genius, must be limited by the conditions that surround him; and science in general was not sufficiently advanced two hundred years ago to be much help to Newton, so that—for lack of information which is ordinary knowledge to us living in the 20th century—he fell into the error of attributing the effects of " weight " and " force " to a common cause, which—for want of a better term—he called gravitation; but I have not the slightest doubt that if he were living now he would have arrived at the following more reasonable conclusions:—That terrestrial things fall to earth by " gravis," weight; because they are matter; while the heavenly bodies (which also are matter) do not fall because they are maintained in their courses by magnetic or electric force.

Another figure of great prominence in the early part of the eighteenth century was Dr. Halley, who survived Sir Isaac Newton by some fifteen years, and it is to him that we owe nearly all the methods of measuring distance which are used in astronomy at the present day. So far no one had seriously considered the possibility of measuring the distance to the sun planets or stars since Hipparchus had failed—away back in the second century B.C.—but now, since the science had made great strides, it occurred to

Dr. Halley that it might be possible at least to find the distance from the earth to the sun, or to the nearest planet.

Remembering the time-honoured dogma that the stars are infinitely distant, inspired by the magnificence of the Copernican conception of the universe, and influenced—no doubt—by the colossal suggestions of Ole Roemer, he tried to invent some means of making a triangulation on a gigantic scale, with a base-line of hitherto unknown dimensions.

Long years ago Kepler had worked out a theory of the distances of the planets with relation to each other, the principle of which—when expressed in simple language and in round figures—is as follows :—" If we knew the distance to any one of the planets we could use that measurement as a basis from which to estimate the others.   Thus Venus is apparently about twice as far from the sun as Mercury, while the earth is about three times and Mars four times as far from the sun as Mercury, so that should the distance of the smallest planet be—let us say—50 million miles, then Venus would be 100, the Earth 150, and Mars 200 millions of miles."

This seems to be the simplest kind of arithmetic, but the whole of the theory of relative distance goes to pieces because Kepler had not the slightest idea of the linear distance from the earth to anything in the firmament, and based all his calculations on time, and on the apparent movements of the planets in azimuth, that is—to right or left of the observer, and to the right or left of the sun.

Necessity compels me to state these facts in this plain and almost brutal fashion, but it is my sincere hope that no reader will suppose that I under-estimate the genius or the worth of such men as Newton and Kepler ; for it is probable that I appreciate and honour them more than do most of those who blindly worship them with less understanding.   I only regret that they were too ready to accept Copernican astronomy as though it were an axiom, and did not

put it to the proof ; and that, as a consequence, their
fine intelligence and industry should have been devoted to the glorification of a blunder.

Kepler's work was of that high order which only
one man in a miliion could do, but nevertheless, his
calculations of the relative distances of the planets
depends entirely upon the question whether they
revolve round the sun or not ; and that we shall
discover in due course.

However, Dr. Halley had these theories in mind
when he proposed to measure the distance to Mars
at a time when the planet reached its nearest point
to earth (in opposition to the sun), and then to multiply
that distance by three (approximate), and in that
manner estimate the distance of the sun. He proceeded then to invent what is now known as the
" Diurnal Method of Measurement by Parallax," which
he described in detail in the form of a lecture to
contemporary astronomers, introducing it by remark-
ing that he would probably not be living when next
Mars came into the required position, but others
might at that time put the method into practice.

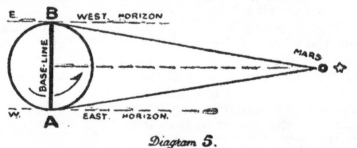

*Diagram 5.*

He began by saying that " If it were possible to
place two observers at points diametrically opposite
to each other on the surface of the earth (as A and B
in diagram 5), both observers—looking along their
respective horizons—would see Mars at the same time,
the planet being between them, to the east of one
observer and to the westward of the other. In
these circumstances the diameter of the earth might

be used as a base-line, the observers at A and B might take simultaneous observations, and the two angles obtained, on being referred to the base-line, would give the distance of the planet."

But this was in the reign of George II. long before the invention of steamships, cables or telegraphs, and Dr. Halley knew that it was practically impossible to have B taking observations in the middle of the Pacific Ocean, so he proposed to overcome the difficulty by the following expedient :—He suggested that both the observations could be taken by a single observer, using the same observatory, thus—" Let an observer at A take the first observation in the evening, when Mars will be to his east : let him then wait twelve hours, during which time the rotation of the earth will have carried him round to B. He may then take his second observation, Mars being at this time to his west, and the two angles thus obtained—on being referred to the base-line—will give the distance of the planet."

This proposition is so plausible that it has apparently deceived every astronomer from that day to this, and it might even now deceive the reader himself were it not that he knows I have some good reason for describing it here.   It is marvellously specious ;  it does not seem to call for our examination ;  and yet it is all wrong! and Dr. Halley has a world of facts against him.

He is at fault in his premises, for if the planet was visible to one of the observers it must be above his horizon, and, therefore, could not be seen at the same time by the other ;  since it could not be above his horizon also.  (See diagram 5.)

Again, his premises are in conflict with Euclid, because he supposes Mars to be midway between A and B, that is between their two horizons, which are parallel lines 8,000 miles apart throughout their entire length, and so it is obvious that if the planet—much smaller than the earth—was really in that position it could not be seen by either of the observers.

The alternative which Dr. Halley proposes is as

fallacious as his premises, for he overlooks the fact
that—according to Copernican astronomy—during the
twelve hours while the earth has been rotating on its
axis it has also travelled an immense distance in its
orbit round the sun.   The results are :—

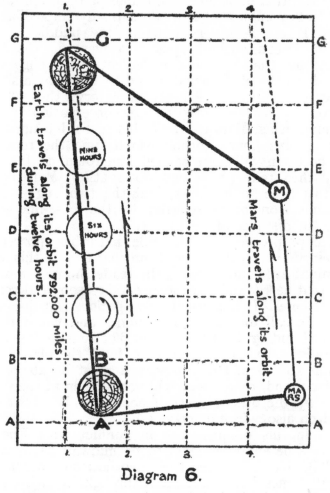

Diagram 6.

1. That an observer starting from A can never
   arrive at B, but must arrive in twelve hours'
   time at a point somewhere about three-quarters of
   a million miles beyond it, as shown in diagram 6.

2. The observer loses his original base-line, which was the diameter of the earth, and does not know the length of his new one, A, G, because the distance of the sun and the dimensions of the orbit had never previously been measured.

3. The angle of view from G is entirely different from the one intended from B.

4. Mars itself has moved along its orbit during the twelve hours, to a new position which is very uncertain.

5. The triangulation which was intended is utterly lost, and the combined movements of the earth and Mars, plus the two lines of sight, make up a quadrilateral figure, which of course contains angles of 360 degrees, and by means of which no measurement whatever is possible.

In conclusion, Dr. Halley was mistaken when he supposed that two observations made from a single station with an interval of twelve hours between them, were equivalent to two observations taken simultaneously by A and B. . .

The actual attempt to measure the distance to Mars by the use of this Diurnal Method will be dealt with in the proper order of events, but for the present—what more need I say concerning such ingenious expedients?

.    .    .    .    .    .

A curious example of theorising to no useful purpose is the "Theory of the Aberration of Light," which is regarded by some as one of the pillars of astronomy. It aims to show that if the velocity of the earth were known the velocity of light could be found, while at the same time it implies the reverse :—that if the velocity of light were known we could find at what speed the earth is travelling round the sun. If Bradley intended to prove anything by this theory it was that the apparent movement of the stars proves that the earth is in motion ; which surely is begging the question.

The fact that the theory of the Aberration of Light has no scientific value whatever is very well shown by the following quotation from its author :—" If the observer be stationary at B (see dia. 7) the star will appear to be in the direction B, S ; if, however, he traverses the line B A in the same time as light passes from the star to his eye the star will appear in the direction A. S."

That is true, but it would be no less true if the star itself had moved to the right while the observer remained at B, but why did he say "if he moves from B to A in the same time as it takes light to pass from the star to his eye"? It is a needless qualification, for if the observer moves to A he will see the star at the same angle whether he walks there at three miles an hour or goes there by aeroplane at a mile a minute. It has nothing to do with the speed of light, and the velocity of light has nothing to do with the direction of the star, it is merely posing, using words to no purpose.

## Chapter Five

## THE DISTANCE TO THE MOON

LET us pass on to something more important, the measurement of the distance to the moon, the first of the heavenly bodies to be measured. This was performed by Lalande and Lacaille in the year 1752, using the method of direct triangulation. Lalande took one of the observations at Berlin, while Lacaille took the other at the same time at the Cape of Good Hope ; a straight line (or chord) joining these two places giving them a base-line more than 5,000 miles in length.

The moon was at a low altitude away in the west, the two observers took the angles with extreme care, and at a later date they met, compared notes, and made the necessary calculations. As a result the moon was said to be 238,830 miles from the earth, and to be 2,159.8 miles in diameter, the size being estimated from its distance ; and these are the figures accepted in astronomy the world over at the present day.

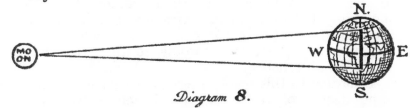

*Diagram* **8.**

I have occasion to call the reader's attention to the fact that some books—Proctor's " Old and New Astronomy " for example—in describing the principle of how to measure to the moon, illustrate it by a diagram which differs from our diagram 8. Though the principle as it is explained in those books seems plausible enough, it would be impossible in practice, for the diagram they use clearly shows the moon to be near the zenith. Further, it is often said that the distance to the moon has been several times measured, but the fact is that it is of no consequence whether it has or not, for it is the result obtained by Lalande and Lacaille which is accepted by astronomy, and their observations were taken as I have stated, and illustrated in diagram 8. Moreover, one of the greatest living authorities on astronomy tells us that their work was done with such precision that " the distance of the moon is positively settled, and is known with greater accuracy than is the length of any street in Paris." Nevertheless we will submit it to the test.

There is every reason to believe that the practical work of these two Frenchmen was most admirably done, and yet their labours were reduced to naught, and the whole object of the triangulation was defeated, because, in making the final computations they made " allowances " in order to conform to certain of the established false theories of astronomy.

One of these is the Theory of Atmospheric Refraction, which would have us believe that when we see the sun (or moon) low down on the horizon, at sunrise or sunset, it is not really the sun itself that we see, but only an image or mirage of the sun reflected up to the horizon by atmospheric refraction, the real sun being at the time at the extremity of a line drawn through the centre of the earth, 4,000 miles below our horizon. (That is according to the astronomy taught in all schools.)

According to this theory there is at nearly all times some degree of refraction, which varies with the altitude of the body under observation, so that (in simple) the theory declares that the real moon was considerably lower than the moon which Lalande and Lacaille actually saw, for that was only a refracted image.

They had, therefore, to make an allowance for

atmospheric refraction.   They had to find (by theory) where the real moon would be, and then they had to modify the angles they had obtained in practical triangulation, by making an allowance for what is known as " Equatorial Parallax."

I will explain it:—Equatorial Parallax is defined as " the apparent change in the direction of a body when seen from the surface of the earth as compared with the direction it would appear to be in if seen from the centre of the earth."

It is difficult not to laugh at theories such as these, but I can assure the reader that astronomers take them quite seriously.   If we interpret this rightly, it is suggested that if Lalande and Lacaille will imagine themselves to be located in the centre of the earth they will perceive the moon to be at a lower altitude than it appeared to them when they saw it from the outside of the earth ;   and   modern   Copernican astronomy required that on their return to Paris they should make allowance for this.

Now observe the result.   It has been shown that " Equatorial Parallax " is only   concerned   with altitude ;   it is a question of higher or lower ; it has to do with observations taken from the top of the earth compared with others taken theoretically from the centre. Really it is an imaginary triangulation, where the line E P in diagram 9 becomes a base-line. The line E P is vertical ; therefore it follows that the theoretical triangulation by which Equatorial Parallax is found is in the vertical plane. . .   We remember, however, that the moon was away in the west when seen by Lalande and Lacaille, while their base-line was the chord (a straight line running north and south) connecting Berlin with the Cape of Good Hope.   These facts prove their triangulation to have been in azimuth ; that is, in the horizontal—or nearly

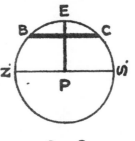

Dia. 9.

horizontal—plane ; indicated by the base-line B C in diagram 9.

Now the three lines of any and every triangle are of necessity in the same plane, and so it follows that every calculation or allowance must also be in that plane ; but we find that while Lalande and Lacaille's triangulation to the moon was in the horizontal plane B C, the allowance they made for Atmospheric Refraction and Equatorial Parallax was in the contrary vertical plane E P ! . . .

By that almost inconceivable blunder real and imaginary angles came into conflict on two different planes, so the triangulation was entirely lost ; and as a consequence the distance of the moon is no more known to-day than it was at the time of the flood.

N.B.—All other attempts to measure the distance to the moon since that time have been defeated in a similar manner.

## Chapter Six

## ROMANTIC THEORIES

THIS history of the evolution of astronomy would not be complete if we omitted to mention here the fact that, though the French school of astronomers had been foremost in adopting practical triangulation, it was not until the British took up the work in 1783 that the triangulation of the earth was seriously begun.

. . . . . .

At about this time Immanuel Kant was laying the foundation of the Nebular Hypothesis—the theory that the earth and the planets were created by the sun.

Sir William Herschell became interested, and carried the thought further, but the Nebular Hypothesis may be said to have been still only in a nebulous state until it was taken up and developed by the brilliant French mathematician and astronomer the Marquis de Laplace.

According to this hypothesis there was a time, ages ago, when there was neither earth, nor moon, nor planets, but only an immense mass of incandescent nebulous matter (where the sun is now), spinning and flaming like a gigantic catherine wheel . . . alone amid the stars.

In other words there was only the sun, much larger than it is at the present time. This mass cooled and contracted, leaving a ring of tenuous blazing matter like a ring of smoke around it. In the course of time this ring formed itself into a solid ball, cooled, and became the planet Neptune.

The sun contracted again, leaving another ring, which formed itself into a ball and became the planet

Uranus, and so it went on until Saturn, Jupiter, Mars, and then the Earth itself were created in a similar way ; to be followed later by Venus and Mercury.

In this way Laplace explained how the earth and the planets came to be racing round the sun in the manner described by Copernicus ; and, strange to say, this Nebular Hypothesis is now taught in the schools of the twentieth century with all the assurance that belongs to a scientific fact.

Yet the whole thing contradicts itself, for the laws of dynamics show that if the sun contracted it would rotate more rapidly, and if it rotated more rapidly that would increase the heat, and so cause the mass to expand.

It appears then, that as every attempt to cool increases the rotation, and heat, and so causes further expansion, the sun must always remain as it is. It cannot get cooler or hotter ! and it cannot grow bigger or less ! and so it is evident that it never could leave the smoke-like rings which Laplace imagined. Therefore we know that the earth could never have been formed in that way; and never was part of the sun.

This Nebular Hypothesis is pure imagination, and it is probable that it was only allowed to survive because it made an attempt to justify the impossible solar system of modern astronomy. It ends in smoke.

.    .    .    .    .    .

Just like a weed—which is always prolific—the Nebular Hypothesis soon produced another equally unscientific concept, known as the Atomic Theory.

The idea that everything that exists consists of— or can be reduced to—atoms, was discussed by Anaxagoras and Democritus, away back in the days of Ancient Greece, but it was not until the beginning of the 19th century that it was made to account for the creation of the entire universe. Let us dissect it.

An atom is " the smallest conceivable particle of matter," that is—smaller than the eye can see, even with the aid of a microscope ; it is the smallest thing the mind of man can imagine. And the Atomic Theory suggests that once upon a time (a long way further back than Laplace thought of) there was nothing to be seen anywhere, in fact there seemed to be nothing at all but everlasting empty space ; and yet that space was full of atoms smaller than the eye could see, and in some manner, which no one has been able to explain, these invisible atoms whirled themselves into the wonderful universe we now see around us.

But if there had ever been a time when the whole of space was filled with atoms, and nothing else but atoms in a state of unity, they must have been without motion ; and being without motion, so they would have remained for ever ! . . . Of course the idea that all the elements could have existed in that uniform atomic state is preposterous, and shows the whole theory to be fundamentally unsound, but if—for the sake of argument—we allow the assumption to stand, the atomic condition goes crash against Newton's " Laws of Motion," which show that " every thing persists in a state of rest until it is affected by some other thing outside itself."

The tide of events now carries us along to the year 1824, when Encke made the first serious attempt to find the distance to the sun ; using as the means— the Transit of Venus.

He did not take the required observations himself, but made a careful examination of the records which had been made at the transits of 1761 and 1769, and estimated the sun's distance from these ; employing the method advocated by Dr. Halley.

What is meant by the " Transit of Venus " is the fact of the planet passing between the observer and the sun (in daylight) when, by using coloured or smoked glasses to protect the eyes, it may be seen as a small spot moving across the face of the solar disk. The method of finding the distance to the sun, at such a time, is as follows:—Two observers are to be placed as far apart as possible on the earth, as B and S in diagram 10. From these positions B will see Venus cross the face of the sun along the dotted line 2, while

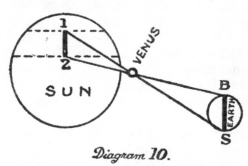

*Diagram 10.*

S will see the planet projected nearer to the top edge of the sun, moving along the line 1. The distance which separates the two projections of Venus against the solar disk, indi-cated by the short vertical line 1—2 will bear a certain proportionate relation to the base-line—or diameter of the earth—which separates the observers B and S.

On referring to the Third Law of Kepler, laid down in the 17th century—it is calculated that the ratio of the line 1—2 as compared with the line B—S will be as 100 is to 37. Consequently, if we know the dimensions of the triangle from B and S to Venus it is a simple matter to find the dimensions of the triangle from Venus to the points 1—2 by the formula—" as 100 is to 37." Further, when we have found the number of miles that are represented by the space which separates the two dotted lines on the face of the sun, we can use the line 1—2 as though it were a yard-stick or a rule, and so measure the size of the sun from top to bottom.

Such is the method which Encke used in his study of the records of transits of Venus which had been made fifty years before, and it is stated on the most reliable authority that the results he obtained were accepted without question.

In round figures he made the sun to be about 97,000,000 miles from the earth and 880,000 miles from top to bottom. All this seems reasonable enough, and it certainly is ingenious ; and yet—

The observers were not—as a matter of fact—placed at the poles, nor were they diametrically opposite to each other as in the diagram, but they observed the Transit of Venus from two other points not so favourably placed, and so "allowances" had to be made in order to find what the dimensions of the triangle B S Venus would have been if the observers had been there to see the transit. . . And in making these allowances our astronomers were all unconscious of the fact that if the observers really had been there (as in the diagram, and as illustrated in all books and lectures on the subject) they could not both have seen Venus at the same time, because A and B are upside down with respect to each other—their two horizons are opposite and parallel to each other—and the planet could not be above the two horizons at the same time. But the allowances were made, nevertheless, and the triangle, which, as we see, was more metaphysical than real, was referred to the Third Law of Kepler ; which had been designed to fit a theory of the solar system which, so far, has not been supported by a single fact. The result of the entire proceeding was " nil."

*Chapter Seven*

# A GALAXY OF BLUNDERS

·THE world of astronomy being satisfied that Encke had really found the distance of the sun, the time had come when a triangulation to the stars might be attempted ; and this was done by F. W. Bessel in the year 1838. He is said to have been the first man to make a successful measurement of stellar distance when he estimated the star known as " 61 Cygni " to be 10½ light-years, or 63,000,000,000,000 miles from the earth ; its angle of parallax being 0.31 '' ; and for this work Bessel is regarded as virtually the creator of Modern Astronomy of Precision.

The reader who has followed me thus far will suppose that I intend to examine this measurement of " 61 Cygni." That is so ; but as it will be necessary to introduce astronomical terms and theories which will be unfamiliar to the layman, I must explain these at some length in order that he, as one of the jury, may be able to arrive at a just verdict. In the meantime I respectfully call the attention of the responsible authorities of astronomy to this chapter, for it is probable that I shall here shatter some of their most cherished theories, and complete the overthrow of the Copernican astronomy they represent.

Light is said to travel at a speed of 186,414 miles a second ; that is 671,090,400 miles in an hour, or six billion (six million millions) miles in a year. So when " 61 Cygni " is said to be 10½ light-years distant it means that it is so far away that it takes its light ten and a half years to travel from the star to the eye of the observer, though it is coming at the rate of 671,090,400 miles an hour. One light-year equals 6,000,000,000,000 miles.

An " angle of parallax " is the angle at the star, or
at the apex of an astronomer's triangulation.    The
angle of parallax 0.31" (thirty-one hundredths of a
second of arc) is so extremely small that it represents
only one 11,613th part of a degree.    There is in
Greenwich Observatory an instrument which has a
vernier six feet in diameter, one of the largest in the
world.    A degree on this vernier measures about three-
quarters of an inch, so that if we tried to measure
the parallax 0.31" on that vernier we should find it
to be one 15,484th part of an inch.    When angles are
as fine as this we are inclined to agree with Tycho
Brahe when he said that " Angles of Parallax exist
only in the minds of the observers ; they are due to
instrumental and personal errors."

The Bi-annual (or semi-annual) method of stellar
measurement which Bessel used for his triangulation
is very interesting, and, curiously enough, it is another
of those singularly plausible inventions advocated by
Dr. Halley.

It will be remembered how Hipparchus failed to get
an angle to the stars 2,000 years ago, and arrived at
the conclusion that they must be infinitely distant ;
and we have seen how that hypothesis has been
handed down to us through all the centuries without
question, so we can understand how Dr. Halley was
led to design his method of finding stellar distance on
a corresponding, in-
finitely distant scale.

It appeared to him
that no base-line on
earth (not even its dia-
meter) would be of
any use for such an
immense triangulation
as the stars required,
but he thought it might
be possible to obtain a

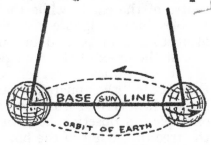

Diagram 11.

base-line long enough if we knew the distance of the
sun; and his reasoning ran as follows :—As we have

learned from Copernicus that the earth travels completely round the sun once in a year, it must be on
opposite sides of the orbit every six months, therefore,
if we make an observation to a star—let us say—to-
night, and another observation to the same star when
we are on the other side of the orbit in six months'
time, we can use the entire diameter of the orbit as a
base-line.            *wrong if axed, & sun = wrong.*

Of course this suggestion could not be put into
practice until the distance to the sun was found, but
now that Encke had done that, and found it to be about
97,000,000 miles, Bessel had only to multiply that by
two to find the diameter of the orbit, so that the length
of his base-line would be, roughly, 194,000,000 miles.
It seemed a simple matter, then, to make two observations to find the angle at the star " 61 Cygni," and to
multiply it into the length of the base-line just as a
surveyor might do.

A critical reader might observe that as there is in
reality only one earth, and not two, as it appears in
diagram 11, the base-line is a very intangible thing
to refer any angles to ; and he might think it
impossible to know what angles the lines of sight really
do subtend to this imaginary base-line ; but these
questions do not seriously concern the astronomer
because the " Theory of Perpendicularity " assures
him that the star is at all times perpendicular to the
centre of the earth, while the " Theory of Parallax "
enables him to ignore the direction of his base-line
altogether, and to find his angle—not at the base !
but at the apex of the triangle—at the star.

These theories, however, deserve our attention ;
Parallax is " the apparent change in the direction of a
body when viewed from two different points." For
example, an observer at A in diagram 12, would see
the tree to the left of the house, but if he crosses over
to B, the tree will appear to have moved to the right
of the house. Now in modern astronomy the stars
are supposed to be fixed, just as we know the tree
and the house to be, and an astronomer's angle of

parallax is " the apparent change in the direction of a star as compared with another star, when both are viewed from two different points, such as the opposite sides of the orbit." The "Theory of Parallax " as stated in astronomy, is " that the nearer the star the greater the parallax; hence the greater the apparent displacement the

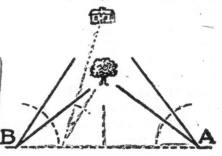

*Diagram 12.*

nearer the body or star must be." In other words, it is supposed that because the tree in the diagram is nearer to the observer than the house, it will appear to move further from the house than the house will appear to move away from the tree, if the observer views them alternately from A and B. That is the principle which Bessel relied upon to find the parallax of " 61 Cygni." (I will leave the reader to make his own comments upon it.)

*Diagram 13.*

The " Theory of Perpendicularity" tells us that all stars are perpendicular to the centre of the earth, no matter what direction they may appear to be in as we see them from different points on the surface; and proves it by "Geocentric Parallax." . . If that is so, then every two observations to a star must be parallel to each other, the two angles at the base must inevitably equal 180 degrees, and consequently there can be no angle whatever at the star ! But the word perpendicular is a relative term. It has no meaning unless it is referred to a line at right angles. Moreover, no thing can be said to be perpendicular to a point ; and the centre of the earth is a point as defined by Euclid, without length, breadth or thickness ; yet this theory supposes a myriad stars all to be perpendicular to the same point. The thing is false. The fact is that the stars diverge in all directions from the centre of the earth, and from every point of observation on the surface. (See diagram 13.) It would be as reasonable to say that all the spokes of a wheel are perpendicular to the hub.

    .    .    .    .    .    .

"So much for the theories ; but Bessel believed in them, because they are among the tenets of astronomical faith ; and he discovered that " 61 Cygni " appeared to move by an 11,613th part of a degree, as compared with another star adjacent to it. So he deduced the parallax 0.31" as the angle of " 61 Cygni," the other star (the star of reference) being presumed to be so much further away as to have no angle whatever.

It appears that—in spite of the fact that the theory of Perpendicularity makes it impossible to obtain any angle to a star—Bessel is supposed to have found an angle by means of parallax ; for although the two lines of sight are as nearly parallel as possible, the parallax 0.31" indicates that they are really believed to converge by that hair's-breadth. Unfortunately for this idea, however, the theory of Perpendicularity is supported by another theory—that of Geocentric Parallax, which makes every line of sight taken at the surface of the earth absolutely parallel to a line from the centre of the earth to the star, wherefore astronomy has the choice of two alternatives, viz. : if these two theories are right, neither Bessel nor anyone else could ever get an angle at the star ; while, on the other hand, if he did obtain an angle,—then the two theories are wrong. Still we have not done with this matter, for the triangulation was made still further impossible by the use of Sidereal Time.

Hipparchus had observed that whereas the sun crossed the meridian every 24 hours, the stars came round in turn and crossed in a little less, so that, for example, Orion would cross the meridian every 23 hours 56 minutes 4.09 seconds. This is called a Stellar or Sidereal day. It is divided into 24 equal parts, or hours, each a few seconds less than the ordinary hour of 60 minutes which is taken by the sun, and it is this Sidereal Time which is used by all modern astronomers, their clocks being regulated to go faster than the ordinary clock, so as to keep pace with the stars as they pass. As Sidereal time is designed to bring every star back exactly on the meridian every 24 hours by the sidereal clock, it follows of necessity that the stars re-appear on the meridian with perfect regularity; (if they do not the clock is altered slightly to make them do so.) The agreement between the star and the sidereal clock becomes a truism, and a law invincible. It is certain, therefore, that if " 61 Cygni " did not appear to be exactly in its appointed place by the astronomer's time, the clock was wrong.

We have now two theories and the sidereal clock to prove that every line of sight to " 61 Cygni " is parallel to every other; that they cannot possibly converge, and consequently that no triangulation was obtained. Let us illustrate it in a diagram: 14. An

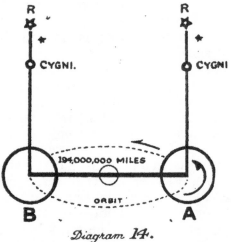

*Diagram 14.*

observer at A sees the star 61 Cygni, and also R, the star of reference; both on his meridian. The earth is supposed to be moving round the sun in the direction of the arrow, until in 182 or 183 sidereal days the observer is at B, and then sees both the stars on his meridian exactly as he saw them before. The two meridians and lines of sight are parallel, so that if continued for ever they can never meet at a point, and the two angles at the base equal 180 degrees, yet the stars are on both lines.

It is obvious, therefore, that the stars have moved to the left (east), precisely as much as the earth has moved to the left in its orbit. If the earth has moved, so have the stars; that is clear. We have proved that Bessel did not get a triangulation to " 61 Cygni," because it is impossible to do so by the semi-annual method; and that the apparent displacement, or parallax 0.31″ was due to error. No such displacement could be discovered unless the clock was wrong, or unless Cygni itself had moved in reality, more or less than the star of reference; wherefore, as every astronomer since 1838 has used the same method, it follows that no triangulation to a star has ever been successfully made; and that every stellar distance given in the modern text-books on astronomy is hopelessly wrong.

Though my case is now really won, and students of astronomy wlll see the justice of my conclusions, this chapter may not be quite complete without the following comments with reference to diagram 14 :—

Reasoning entirely from the standpoint of the Copernican Theories, we have seen that if the earth has moved from one side of the sun to the other (from A to B), so also have the stars ; but astronomers know as well as I do that the stars do not move eastward, neither do they—in nature—even appear to do so ; their movement (real or apparent) being beyond all doubt—to the westward. So it is established that the stars have not moved eastward from A to B, and this—added to the fact that they really would be in the same positions with respect to the meridian as shown in the diagram, proves that the earth has not moved eastward either. And as

the earth has not moved from A to B, as Dr. Halley
and Bessel believed, the base-line disappears, the orbit
no longer exists ; and with the orbit falls the whole
solar system of Nicholas Copernicus.

   N.B.—If the earth remained at A rotating on its
       axis once in every sidereal day, the stars
       would appear always as shown at A—on the
       meridian at the end of every revolution ;
       but then we could not account for the fact
       that the sun is on that meridian at the end
       of every solar day—which is nearly four
       minutes longer than the stellar day. On the
       other hand, if we assume the earth to be
       rotating on its axis once in every 24 solar
       hours, we could not then account for the
       stars being on the meridian every 23 hours
       56 minutes 4.09 seconds, as we have proven
       them to be ; and so we arrive at the only
       possible explanation, which is—that the earth
       remains always at A and does not rotate at
       all ; but the sun passes completely round it
       once in 24 hours, while the stars pass round
       it (from east to west) once in every sidereal
       day ; thus they re-appear on the meridian
       at every revolution, including the 183rd ;
       and so we find that the star " Number 61 in
       the Swan " (Cygni) was observed twice from
       the surface of an earth which has never
       moved since the creation. Thus we know
       that the stars are not fixed, as Copernicus
       believed ;    and    the    edifice  of   modern
       astronomy—which Sir Robert Ball described
       as " the most perfect of the sciences " might
       be more truly described as the most amazing
       of all blunders.

## Chapter Eight

# MARS

IDEAS that have been familiar to us from our very
earliest childhood, which we have heard echoed on
every hand, and seen reflected in a thousand ways,
are tremendously hard to shake. We seem to love
them as part of ourselves, and cling to them in the
face of the most overwhelming evidence to the
contrary.

So it often happens that men and women whose
common sense and reason tells them that many of the
statements of astronomy are as incredible as the
story of Jack and the Beanstalk, are still loth to part
with their life-long beliefs, and suggest that, after
all, the modern theory must be true because astronomers
are able to predict eclipses.

But the Chaldeans used to predict the eclipses
three thousand years ago ; with a degree of accuracy
that is only surpassed by seconds in these days
because we have wonderful clocks which they had
not. Yet they had an entirely different theory of
the universe than we have. The fact is that eclipses
occur with a certain exact regularity just as Christmas
and birthdays do, every so many years, days and
minutes, so that anyone who has the records of the
eclipses of thousands of years can predict them as
well as the best astronomers, without any knowledge
of their cause.

The shadow on the moon at the lunar eclipse is said
to be the shadow of the earth, but this theory received
a rude shock on February 27th, 1877, for it is recorded
in M. Camille Flammarion's " Popular Astronomy "
that an eclipse of the moon was observed at Paris on

that date in these circumstances : " the moon rose at 5.29, the sun set at 5.39, and the total eclipse of the moon began before the sun had set."

The reader will perceive that as the sun and moon were both visible above the horizon at the same time for ten minutes before sunset, the shadow on the moon could not be cast by the earth. (See diagram 31.) Camille Flammarion, however, offers the follow-

31. The Eclipse.

ing explanation : He says, " This is an appearance merely due to refraction. The sun, already below the horizon, is raised by refraction, and remains visible to us. It is the same with the moon, which has not yet really risen when it seems to have already done so."

Here is a case where modern astronomy expects us to discredit the evidence of our own senses, but to believe instead their impossible theories. . . This Atmospheric Refraction is supposed to work both ways, and defy all laws. It is supposed to throw up an image of the sun in the west—where the atmosphere is warm, and at the same time to throw up an image of the moon in the east—where it is cool ! It is absurd.

. . . . . .

When speaking of the measurement of the distance to Mars by Sir David Gill, in the same year, 1877, Sir Norman Lockyer described it as " One of the noblest achievements in Astronomy, upon which depends the distance to and the dimensions of everything in the firmament except the moon." Evidently a very big thing, worthy of our best attention. The method which Sir David Gill used was the " Diurnal

Method of Measurement by Parallax," which we have dealt with in an earlier chapter. He adopted the suggestion made by Dr. Halley, and took the two observations to Mars himself, at Ascension Island, in the Gulf of Guinea.

The prime object of the expedition was really to find the distance to the sun (though we remember that that had been done by Encke fifty years before by the Transit of Venus), which was to be done by first measuring the distance to Mars, and, having found that, by multiplying the result by 2.6571 (roughly 3), as suggested by Kepler's Theory of the relative distances of the sun, earth and planets, in this manner : Distance to Mars, 35,000,000×2.6571 = 93,000,000 miles.

The Encyclopædia Britannica tells us that " The sun's distance is the indispensable link which connects terrestrial measures with all celestial ones, those of the moon alone excepted, hence the exceptional pains taken to determine it," and assures us later that " The first really adequate determinations of solar parallax were those of Sir David Gill—result 8.80"," and that his measures " have never been superseded."

He found the Angle of parallax of Mars to be about 23", which made its distance to be 35 million miles, and this, multiplied by 2.6571, showed the sun to be 93 million miles in the opposite direction. We realize that although the sun's distance is said to be the indispensable link, it depends upon the measurement to Mars, so that this is more indispensable still. It is the key to all the marvellous figures of astronomy, and for that reason we will give it special treatment.

The figure 35,000,000 miles depends upon the angle at the planet, which is an angle of parallax. That is— the apparent change in the direction of Mars to the right or left of the star x (star of reference) when both are viewed from the opposite ends of a base-line, which, in this case, is the diameter of the earth ; see diagram 15. Theory : If Mars is much nearer than x, and both are on a line perpendicular to the centre of the earth, an observer at A will see the planet to the left or east of the star, while B will see it to the right or west of that star. (East and west are local terms, and change with the position of the observer.)

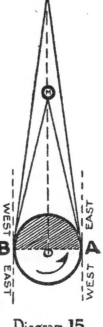

The star of reference is presumed to be billions of miles away, so far away, indeed, that it is supposed to have no angle at all, so that the lines A x and B x are really parallel to each other, and at right angles to the base-line, as shown in diagram 16. Even Mars is at a tremendous distance, so that the angle of parallax is the very small fraction of a degree by which the planet is less perpendicular than the star. Nevertheless, however slight the apparent displacement of Mars may be, if it is be-

Diagram 15.

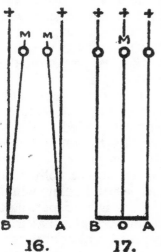

16. 17.

tween the two perpendiculars A x, and B x, the lines of sight A M and B M would meet somewhere at a point.

So far we have supposed A and B to be making observations at the same time, but Sir David Gill believed with Dr. Halley that he might take the two observations himself, the first from A in the evening, and the second from B the next morning, allowing the rotation of the earth to carry him round from A to B during the night, and that these two observations would give the same result as two observations taken by A and B at the same Greenwich time. Accordingly he took two observations at Ascension Island, one to his east and the other to the west, and, relying upon all the theories of his predecessors, failed to perceive that his second line of sight to the planet was on the wrong side of the perpendicular, and diverged from the first.

The fundamental principle of parallactic angles is unsound, while it is at the same time in conflict with quite a host of other astronomical theories, because the theories of Atmospheric Refraction, Perpendicularity, Geocentric Parallax, and the Aberration of Light, combined with the use of Sidereal Time, all go to prove that every observation taken from the surface of the earth to a star is exactly parallel with a line from the centre of the earth to the same star, and that B's line to x is parallel to that of A.

Consequently if Mars were on the line O X (in diagram 15), as Dr. Halley presumed when he invented this method, it would be perpendicular to both A and B, therefore neither one observer or the other would see it at any angle at all ; as shown in diagram 17. It is not possible for any observer on earth to see Mars to the right or left of a star that is perpendicular unless the planet is in reality to the right or left of that perpendicular. No apparent displacement could occur, but the displacement must be physical ; and so the theory of parallactic angles is exploded.

Of course there will be some ready to contend that Sir David Gill really did measure an angle. That is true ; but it will prove to be an actual (physical) deviation of the planet from the perpendicular, which is a very different thing than an angle of parallax. But it was believed to be a parallactic angle, that is to say—it was supposed to be only an optical or apparent displacement due to the change in the position of the observer from A to B, hence a world of romance is built upon that little angle in this fashion : Angle of Mars 23″ = 35,000,000 miles, $\therefore$ 35,000,000 × 2.6571 = 93,000,000 = solar parallax 8.80″ = distance of the sun ∴ the sun's diameter is 875,000 miles ; weight XYZ lbs., age 17,000,000 years, and will probably be burnt out in another 17 million years. 93,000,000 × 2 = 186,000,000 miles diameter of earth's orbit, the distance to the stars must be billions of miles or even more, they must be a terrific size, and the earth is only like a speck of dust in the Brobdinagian Universe, &c., &c., &c.

But we have not yet done with that angle. Regarded as an angle of parallax, and considered to be equivalent to just such an angle as a surveyor would use in measuring a plot of land, it was of course presumed that the two lines of sight converged so as to meet at a point thirty-five million miles away. (See diagram 18.) This, however, is a mistake, for the two lines of observation, when placed in their proper relations to each other, and in the order as they were taken, should be as in diagram 19, which shows that they diverge.

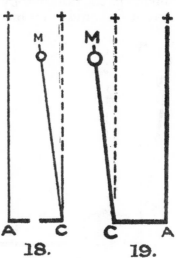

We will prove this in diagram 20. A study of our earlier diagram 6—which gives a suggestion of a small section mapped out with dotted lines to indicate latitude and longitude in universal space—reveals the fact that twelve hours' rotation of the earth does not transfer the observer from A to the point B in space, because—according to Copernican astronomy— the earth is not only rotating on its axis during those twelve hours, but also rushing through space in a gigantic orbit round the sun at the rate of sixty-six thousand miles an hour, or thereabouts, and so when the observer takes his second observation he is something like three-quarters of a million miles away from where he started. He is at latitude G in diagram 6.

Now let us study diagram 20, which has been made as simple as possible in order to illustrate the principles involved the more clearly. The letter C is used in this diagram to take the place of G in the earlier diagram 6, because it is simpler to describe the movements of the observer by A, B, C than it is by A, B, G ; easier to convey my meaning.

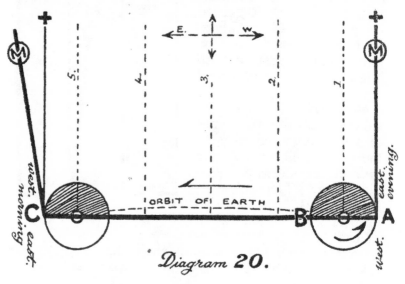

*Diagram* **20.**

All the principles and theories of modern astronomy have been carefully observed, and the parallelism of the lines is strictly in accordance with the theories of Greenwich. As I anticipate that in the course of time a battle-royal will wage around this question of the measurement to Mars, I wish to make it quite clear that diagram 20 is designed only to illustrate the principles ; it is to clarify the whole proceeding so that the layman can follow the argument. If the Royal Astronomical Society have any objection to make, I will be happy to discuss these questions with them in a manner worthy of the subject. The discussion may then, perhaps, be more refined, indeed,

I foresee a very pretty debate, wherefore I advise
them that I know that Sir David did not really take
his observations with a twelve hours' interval as
proposed by Dr. Halley—because it was impossible—
but that he actually waited only seven and a half
hours (hence my use of C in place of G in diagram 20),
but that only elevates the discussion to a higher
plane, while the principle and the net results remain
the same.   In the appointed time and place I will
discuss the actual practice if desired, but here I am
dealing with the principle ;  and talking to the layman
and the judge.

Now let us get on with this diagram 20.   The first
observation is taken at A and the second at C.   It
was evening when the observer was at A, but it is
morning when he arrives at C, so that his east and
west are reversed, the sun remaining fixed far below
the bottom of this page.   (The sun is at the observer's
west in the evening, and to his east in the morning, while Mars is in the opposite direction to the
sun.)

In this example I have placed the planet exactly on
the perpendicular from A to the star of reference,
thus " A MARS X."   That is the starting point, or
first observation ; taken in the evening to the observer's
eastward.   Twelve hours later the observer is at C,
and sees the same star and the planet both to his west ;
but Mars is at this time not exactly on the perpendicular, but a little, a very little, to the left of the star.
The planet is not quite as much west as the star, that
is to say—being to the left—it is to the eastward in
universal geography ;  and to the eastward of the
perpendicular line C X.

Now if we were not particularly careful, and had not
this diagram to guide us, it would be quite natural to
think that the first observation (to the east) should be
on the left hand, and the other (west) on the right,
so as to face each other, so that any angle that might
appear, such as an angle of parallax, would be between
the two perpendiculars to the star.   In that case

they would seem to be as shown in diagram 18 ; but that is wrong !

Referring again to diagram 20, where the observations are illustrated in the proper order as they were actually taken, and all in accordance with the theories of Copernican astronomy, we find that the angle of Mars is to the EASTWARD ! outside of the two perpendiculars. This is more simply shown in diagram 19. A being the first observation, on the right, and C, the second observation, on the left ; that is correct.

Starting, as we did, with Mars on the perpendicular at A, we know that whenever we shall see it again it must be to the eastward of the star which marks that perpendicular, because, while the star remains fixed in space the planet is moving every hour along its orbit to the eastward round the sun, and so, when we see it from C the next morning, it is as we have shown in diagrams 19 and 20. It has moved from the line A X to a position a little further east in universal space than the line C X.

Whatever displacement there is, is outside the two perpendiculars ; so that the second line of sight to Mars diverges from the first ; consequently no triangulation occurs, and nothing of any material value is

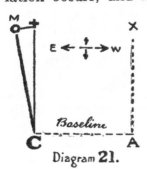

Diagram 21.

accomplished. The so-called angle of parallax was a displacement due to a real movement of the planet during the night.

In conclusion, as A X and C X are one and the same perpendicular, and no angle, either real or apparent, occurs between them, the first observation A X and the base-line are entirely without value, and may be discarded as useless. (Diagram 21.) This leaves us with only the perpendicular C X and the second observation, which proves to be a narrow inverted triangle " C X Mars," where the displacement of the

planet X M—(hitherto known as the parallax of Mars)
—indicates how much the planet has moved to the
left of the star during the night ; while the observer
at C is at the apex.   Just that, and no more.

*Chapter Nine*

# THE TRANSIT OF VENUS, AND THE DISTANCE TO THE SUN.

NOT content with the work already done, all the world of astronomy set out to try to measure the distance to the sun again in the years 1874 and 1882, by observations of the Transit of Venus.

It was a most elaborate affair, 'tis said to be by far the greatest and most costly business ever undertaken for the purposes of astronomy. Men were trained specially for the work, equipped with all the most expensive things in the way of telescopes and instruments, and sent out by the British, French and German governments, all allied for the purpose, as expeditions of astronomers to all parts of the world in order to see Venus—like a small speck—pass across the face of the sun. We have it on the best authority that the 1874 transit was a failure; but, nothing daunted, the expeditions went out again in 1882, to the Indies, the Antipodes and the polar regions, but again the results are admitted to be unsatisfactory; though we may at least hope the astronomers found some entertainment by the way.

The Venus method has already been explained in an earlier chapter, and illustrated in diagram 10. It required that observations should be taken simultaneously by two observers placed as widely apart as possible in order to have the longest base-line obtainable; the ideal base-line being the entire diameter of the earth. From among a large number of observations taken in different parts of the world, two were selected as being better than the rest; they were the observations taken at Bermuda—those lovely little islands near the West Indies—and Sabrina

Land, on the edge of the icy Antarctic regions ; and from this pair the distance of the sun was computed, but the result obtained has never been considered good enough to take the place of the earlier figures of Gill. We will give it the coup-de-grace in short order :—

Bermuda is situated in 32° 15′ north latitude, and 64° 50′ west longitude ; while Sabrina Land is 67° south, and 120° east of Greenwich. We must also mention the fact that both the sun and Venus were somewhere between these places, in the eastern hemisphere.

These common-place facts alone prove that the two observations were not taken at the same time, and consequently were useless for the purpose. I will explain how that is. In their endeavour to secure the longest possible base-line our astronomers separated themselves by 99 degrees in the north and south direction, and by 184° 50′ east and west, so it is perfectly plain that the sun had already set to the observer at Sabrina Land, before the observer at Bermuda could see it rise above his horizon at dawn.

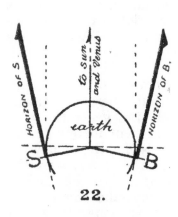

22.

N.B.—The sun rises and sets at a distance of 90 degrees from the observer, so that the Transit astronomers should not have been more than 180 degrees apart even if they had wished to see the sun on the horizon; but our observers had exceeded the limit by nearly five degrees. (See dia. 22.)

The two horizons diverge from each other, and for some part of the time the sun is between them, and not visible to either observer, while as it must be above each of these observer's horizons in turn in order to be seen at all, it is ridiculous to imagine that any observations taken by B and S in a direction toward the top of this page and above their horizons could ever meet anywhere in the universe. The whole business was a fiasco.

Of all the various methods of estimating the distance of the sun, that by means of the measurement to Mars is by far the most important, while the second in order of merit is the one we have just dealt with ; the computation by the transit of Venus, which, it will be remembered, was first used by Encke in 1824. But there are, no doubt, many adherents of astronomy who will still hope to save the time-honoured dogma which hangs upon the question of the distance to the sun ; too egotistical to admit that they could have been mistaken, or too old-fashioned to accept new truths ; and so—while they cannot any longer defend the Mars and Venus illusions—they will say that they know the sun is 93,000,000 miles away because it has been estimated and verified by quite a number of other methods, with always the same result, or thereabouts.

In these circumstances it becomes necessary for us to touch upon these also. The brief examination we shall give to them will be illuminating, and Astronomers will probably be surprised in one way while the layman will be surprised in another. . . . There are some things which every man or woman of ordinary intelligence knows are nonsensical ; but when such things have been permitted to pose for generations as scientific knowledge it is not sufficient merely to say that they are absurd ; they must—for the moment—be treated as seriously as though they really were the scientific concepts they are supposed to be, and it must be shown just how, and why, and where, they are absurd. Then, when that is done, they can masquerade no more, and will no longer obstruct the road to knowledge.

Any one of these means of estimating the sun's distance might be made the subject of a lengthy

argument, for they are like " half-truths " which, as
we all know, are harder to deal with than down-right
falsehood ; but I do not wish to worry the reader
with any more words than I am compelled to use, and
so will deal with them as briefly as possible.

Every one of these things which are believed to be
methods of computing the distance to the sun, or means
of verifying the 93,000,000 mile estimate, presumes the
distance of the sun to be already known ; and in every
case the method is the result of deductions from the
figure " 93,000,000 miles." I am not particularly
concerned as to how or why this was done, nor is it
my affair whether it seems incredible or not : but I
do know that it is as I have stated, and that I am
very well able to prove it. I am only interested in
knowing the truth, and in proving it by reason and
fact.

The verification of the sun's distance by the measurements to the minor planets Victoria, Iris and Sappho,
in 1888 and 1889, was done in the same manner as
the measurement to Mars, and fails in precisely the
same way, by the fallacy of Dr. Halley's Diurnal
Method of Measurement by Parallax.

.    .    .    .    .    .    .

There is the calculation of the sun's distance by the
" Nodes of the Moon," which it is not necessary for
me to dilate upon, because it has already been dis-
credited, and is not considered of any value by the
authorities on astronomy themselves.

.    .    .    .    .    .    .

The computation of the distance to the sun by the
" Aberration of Light " is based upon the theory
that the earth travels along its orbit at the velocity
of 18.64 miles per second. This velocity of the earth
is the speed at which it is supposed to be travelling
along an orbit round the sun, 18.64 miles a second,
66,000 miles an hour, 1,584,000 miles a day, or five
hundred and eighty-four million miles in a year.

The last of these figures is the circumference of the orbit, half of whose diameter—the radius—is of course the distance of the sun itself, and it is from this (pardon the necessary repetition) distance of the sun, first calculated by Encke in 1824, and later by Gill in 1877, that the whole of the figures—including the alleged " velocity of the earth 18.64 miles a second "—were deduced. The 18.64 miles is wrong, because the 93,000,000 is wrong, because neither Encke nor Gill obtained any measurement of the sun's distance whatever; and the whole affair is nothing more than a playful piece of arithmetic, where the distance of the sun is first presumed to be known; from that the Velocity of the earth per second is worked out by simple division, and then the result is worked up again by multiplication to the original figure, " 93,000,000," and the astronomer then says that is the distance to the sun. That is why it is absurd.

.        .        .        .        .        .

The estimation of the distance of the sun by the " Masses of the Planets " depends upon the size, weight, volume or masses of the planets, which depend upon their distance; and the distances of the planets were calculated by Kepler's, Newton's and Bode's Laws from Sir David Gill's attempt to measure the distance of Mars; wherefore, as we have discovered that he did not find the distance to Mars, all the calculations which are founded upon his entirely erroneous conception of the distance, size, and mass of that planet, go by the board.

It will not do for anyone to say to us that the distance to Mars is 35,000,000 miles (when in opposition) and therefore it must be 4,200 miles in diameter, therefore the distance of the sun must be 93,000,000 miles, therefore its diameter must be 875,000 miles and its mass 1,300,000 times greater than the mass of the earth, or three million times greater than Mars, &c., &c., &c., and therefore it must be 93,000,000 miles away. It is neither good logic, good mathematics, nor good sense. If anyone seeks to show that

the distance from the earth to the sun can be measured
by weighing the sun and the planets let him do his
weighing first, and not assume anything; and he
would do well to remember that " The sun's distance
is the indispensable link which connects terrestrial
measures with *all* celestial ones."

Finally the sun's distance as 93,000,000 miles is said
to be justified by the " Velocity of Light." The
Velocity of Light was measured by an arrangement of
wheels and revolving mirrors in the year 1882 at the
Washington Monument, U.S.A., and calculated to be
186,414 miles a second.

> N.B.—Experiments had been made on several
> previous occasions, with somewhat similar
> results, but Professor Newcomb's result
> obtained in 1882, is the accepted figure.

Taking up this figure, astronomers recalled that in
the 17th century Ole Roemer had conceived the
hypothesis that light took nearly $8\frac{1}{4}$ minutes to travel
from the sun to the earth, and so they multiplied his
$8\frac{1}{4}$ minutes by Newcomb's 186,414, and said, in effect
—" there you are again—the distance of the sun is
93,000,000 miles." It is so simple; but we are not
so simple as to believe it, for we have shown in diagram
4 how Ole Roemer deduced that $8\frac{1}{4}$ minute hypothesis from a mistaken idea of the cause of the difference in the times of the Eclipses of Jupiter's Satellites;
and we know that there is no evidence in the world
to show that light takes $8\frac{1}{4}$ minutes to come from the
sun to the earth, so the altogether erroneous and
mis-conceived hypothesis of Ole Roemer can not be
admitted as any kind of evidence and used in conjunction with the calculation of the Velocity of Light
as an argument in favour of the ridiculous idea that
the sun is ninety-three—or any other number of
millions of miles from this world of ours.

All the extraordinary means used by astronomers
have failed to discover the real distance of the sun,

and the many attempts that have been made have achieved no more result than if they had never been done ; that is to say—that it is not to be supposed that they may perhaps be somewhere near the mark ; but it is to be understood, in the most literal sense of the word, that the astronomers of to-day have no more knowledge of the sun's real distance than Adam. Indeed we have to forget all the romantic things that have been said since the time of Copernicus, and look at the universe, as frankly, and as fearlessly as he did : then we might acknowledge the debt we owe to such as he, for even though he was so greatly in error his originality stimulated the world of thought tremendously ; and in that way furthered the world's progress. And then, tutored and encouraged by the shades of Hipparchus, Ptolemy, and Copernicus ; Kepler, Newton and all their kind, we might, with the added experience and advantage of our times, rebuild the science of astronomy as they would do it now ; true to the facts of nature.

## Chapter Ten

# THE BIRTH OF A NEW ASTRONOMY

IT is for me, now, to show how the distance to the sun is really to be ascertained, and this may indicate the way to a new astronomy, and a saner conception of the universe.

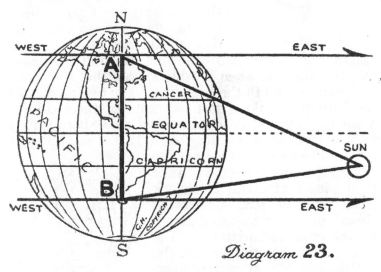

*Diagram* **23.**

¶The Copernican astronomy has been so hedged about with specious theories that it would seem to be impossible to obtain any kind of triangulation to the heavenly bodies that cannot be negatived by Perpendicularity, Geocentric Parallax or similar theories, nevertheless it can be done—and that by two simultaneous observations taken from a base-line which is on solid earth ; thus :—

Let two observers be placed on the same meridian ; A in the northern hemisphere at about Mansfield, Nova Scotia, for example, 60° N, 74° W., and B in the southern hemisphere at Tierra del Fuego, Cape Horn, 55° S. 74° W., as shown in diagram 23. As the two observers are on the same meridian they use the same north and south, while all lines which cross that meridian at right angles indicate east and west, and are parallel to each other; so that A's east is parallel to B's, and to the equator, as in diagram 24. The chord—that is a straight line connecting the two points of observation A, B, will give them a base-line 6,900 miles in length, which runs in a direction due north and south as in diagram 25. The two observers will find their easts by the compass, when it will be seen that they form two right angles to the base-line. The two easts, with the base-line, make a sort of frame, or three sides of a square; and it is within this frame—between the two dotted lines running east, that the triangulation will be made to the sun.

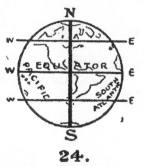

24.

Now let our observers take their places at about 8 o'clock local time (1 p.m. Greenwich Mean Time) on a morning within a week or so of Christmas. The sun will at that time be in the zenith, and almost exactly overhead, at the island of St. Helena, off the coast of South Africa. The observer at A in Nova Scotia will see the sun, blood red, just rising above the horizon to his east-south-east, while the observer at Tierra del Fuego will see the sun at the same time, about eight degrees to the northward of his east (east by north); and so the two lines of sight from A and B converge so as to meet at the sun, which is between the two easts, a little to the southward of A and to the northward of B.

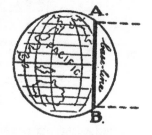

25.

A true triangulation is thus obtained, and the two angles may be referred—either to the parallel easts, or to the base-line which connects them.  No " allowances " of any kind whatever are to be made, and none of the fantastic theories of astronomy are in any way concerned.  It is a plain, ordinary, common-sense triangulation, such as any surveyor would make if we were buying a piece of land ; and that is good enough for us.  The angles at the base-line will equal about 148 degrees, while the angle at the sun, or apex of the triangle, will be 32 degrees (approximate).  When these are multiplied into the base-line by ordinary trigonometry, the sun will prove to be about 13,000 miles in a bee-line from A and 10,000 miles from B.

.        .        .        .        .        .

The stars and the planets are to be measured in a similar manner, when it will be found that no star is at any time further than twenty thousand miles away. As it is my intention to deal more fully with such measurements in another book sequel to this—devoted to the reconstruction, or rather, to the creation of a new Astronomy—I have been content here to say only sufficient to establish my case, and to show that Hipparchus was mistaken when he thought the heavenly bodies were infinitely distant.

And that, truly, is my case, for at last I have shown that the " infinitely distant " hypothesis which has been the guiding star of astronomers for two thousand years, was indeed, an error.

*Chapter Eleven*

# THE EARTH STANDS STILL

IT would seem that Copernican Astronomy had reached its highest development about the year 1882, and then began to decline, or rather, to fall to pieces. The first evidence of this devolution is to be found in the Michelson-Morley experiment of 1887, at Chicago ; the result of which might have undeceived even the most devoted believer in the theory of a spinning earth.

. Professor Michelson was one of the physicists foremost in determining the Velocity of Light, while he has recently been described in the *New York Times* as America's greatest physicist ; and it was he who— working in collaboration with Morley—in 1887 made the most painstaking experiments by means of rays of light for the purpose of testing, verifying, or proving by physical science, what really was the velocity of the earth. To express this more clearly, Astronomers have for a very long time stated that the earth travels round the sun with a speed of more than eighteen miles a second, or sixty-six thousand miles an hour. Without in any way seeking to deny this statement, but really believing it to be thereabouts correct, Michelson and Morley undertook their experiments in order to put it to a practical test ; just in the same way as we might say " The greengrocer has sent us a sack of potatoes which is said to contain 112 pounds weight ; we will weigh it ourselves to see if that is correct."

More technically, the experiment was to test what was the velocity with which the earth moved in its orbit round the sun relative to the æther.

A very well illustrated account of that experiment
will be found in *The Sphere*, published in London,
June 11th, 1921, and it is from that article I quote
the following, verbatim :—" But to the experimenters'
surprise no difference was discernible. The experiment was tried through numerous angles, but the
motion through the æther was NIL ! "

Observe that the means employed represented the
best that modern physical science could do to prove
the movement of the earth through ethereal space,
and the result showed that the earth did not move at
all !  " The motion through the æther was NIL." . . .
But the world of astronomy has not accepted that
result, for it continues to preach the old dogma ; it
appears that they are willing to accept the decisions
of physicists when it suits their case, but reject them
when otherwise. And so they still maintain the
fabulous theory that the earth is rushing through
space at eleven hundred miles a minute ; which, as
they would say in America, "Surely is some traveling."
It must be faster than a bullet from a Lewis gun.

.          .          .          .          .          .

What I have now to record, I do with regret, and
only because my sense of duty in the pursuit of truth
compels me.  It is the circumstance that Sir George
Airy, who retired from his position as Astronomer
Royal in 1881, related—some nine years later—how
he had for some time been harassed by a suspicion
that certain errors had crept into some of the computations published in 1866, and that, though he had
set himself seriously to the work of revision, his powers
were no longer what they had been, and he was never
able to examine sufficiently into the work.  Then he
spoke of a " grievous error that had been committed
in one of the first steps," and pathetically added—
" My spirit in the work was broken, and I have never
heartily proceeded with it since."

My sympathy goes out to Sir George in his tribulation
of the spirit due to advancing age, while I am not

unmindful of myself, for I realize that in him I have lost one who would have been a friend, who would have listened when I said that all was not as it should be with the science of astronomy ; and stood by my side, encouraging and helping, when I, younger and stronger, strove to put it right. I do not know whether Sir George Airy was influenced or not by the result of the Michelson-Morley experiment, but it is at least a noteworthy coincidence that he made those comments only three years later ; but in any case science has need of him, and of such evident open-mindedness and sincerity as his, now.

    .    .    .    .    .    .

Not content to believe that the earth did not move, further experiments were carried out by Nordmeyer in the year 1903, to test the earth's velocity in relation to the Intensities of Light from the heavenly bodies, but he also failed to discover any movement.

    .    .    .    .    .

Even then astronomers were determined to hold on to their ancient theories, and deny the facts which had been twice demonstrated by the best means known to modern physical science. They preferred to believe the theory that the earth was gyrating round the sun with the velocity of a Big Bertha shell, and tried to account for the physicists' failure to discover its movement by finding fault with the æther (or ether). It is not only difficult to understand why they should prefer theory to fact in this manner, and so deceive themselves ; but it is strange also that the world in general could tolerate such nonsense.

However, the results of several years' speculations concerning ether and space were set forth in the year 1911, in a series of lectures by Professor Ormoff, Onspensky and Mingelsky, at Petrograd.

It was suggested that light was not permitted to come from the stars to earth in a straight line, because some quality in ethereal space caused it to follow the earth as it moved round the orbit ; and that might account for the failure of the experiments of 1887 and 1903. In other words it was suggested that we cannot see straight, or that the image of the star as we see it twinkling there is coming to us in a curve— following the earth like a search-light, while it describes the five terrestrial motions ascribed to it by Newton.

When stated even more plainly it means that when we think we see a star overhead we are mistaken, for that is merely the end of a ray of light coming to us from a star which—in the material body—may be millions of miles to the right of us, or it might even be behind us ; as in diagram 26.

Diagram 26.

N.B.—A much greater curvature than we have illustrated in the diagram has since been suggested in all seriousness by leading astronomers from the platform of the R.A.S. at Burlington House, Nov. 6th, 1919, in these words—" . . . . All lines were curved, and if they travelled far enough they would regain the starting point."

Moreover, Ormoff, Onspensky and Mingelsky had come to the conclusion that nothing was fixed in the universe ; so that while the moon goes round the earth and the earth and the planets go round the sun, the sun itself is moving with probably a downward tendency, carrying the whole Copernican solar system with it. Further, even the stars themselves have left their moorings, so that the entire visible universe is drifting ; no one knows where.

In brief, these Petrograd lectures of 1911 introduced many new ideas such as those which have become familiar to the reader in Einstein's Theory of Relativity, since the year following the great World-War.

## Chapter Twelve.

## " RELATIVITY "

THE Theory of Relativity is so complicated, that when it first came to the public notice it was said that there were probably not more than twelve people in the world capable of understanding it. But public interest was aroused, partly by the novelty of Einstein's hypothesis, and partly by the spectacular manner in which it had been received by the British Royal Astronomical Society on the night of November 6th, 1919, until Mr. Eugene Higgins, of U.S.A., offered a prize of 5,000 dollars for the best explanation of relativity, in the form of an essay, describing it so that the general public could understand what it was all about.

The prize was won by Mr. L. Bolton, London ; and his essay can be found in the *Scientific American* (New York and London), June 1921, and also in the *Westminster Gazette*, London, June 14th, 1921. The editor of the *Gazette* found it necessary to remark, when publishing the essay, that " Our readers will probably agree that even when stated in its simplest form it remains a tough proposition."

That is just the trouble with it. It is about as far removed from ordinary " fact " and " plain English " as it is possible for anything to be ; indeed it is so intangible that it may well be that Einstein can form a mental picture of it himself, while he is at the same time unable to convey his meaning to others through the medium of ordinary language.

The thing is elusive ; abounding in inference, suggestion, half-truth and ambiguity ; wherefore it follows that any discussion of it, such as we propose to enter upon, must of necessity be almost equally refined. It might seem tortuous to some readers, and yet be like a very entertaining game of chess to others ; while it certainly will be useful to those who are willing to traverse the long and difficult labyrinth that leads to truth.

Relativity is clever ; but it belongs to the same category as Newton's Law of Gravitation and the Kant-Herschell-Laplace Nebular Hypothesis, in as far as it is a superfine effort of the imagination seeking to maintain an impossible theory of the universe in defiance of every fact against it. . . . Let us see what we can do with it.

First, we will let Professor Einstein himself tell us what he means by Relativity, in the words he used in the opening of his address at Princeton University, U.S.A. :—

" What we mean by relative motion in a general sense is perfectly plain to everyone. If we think of of a waggon moving along a street we know that it is possible to speak of the waggon at rest, and the street in motion, just as well as it is to speak of the waggon in motion and the street at rest. That, however, is a very special part of the ideas involved in the principle of Relativity."

That would be amusing if we read it in a comic paper, or if Mutt and Jeff had said it ; but when Professor Einstein says it in a lecture at the Princeton University, we are expected not to laugh ; that is the only difference. It is silly, but I may not dismiss the matter with that remark, and so I will answer quite seriously that it is only possible for me to speak of the street moving while the waggon remains still—and to believe it—when I cast away all the experience of a lifetime and am no longer able to understand the evidence of my senses ; which is insanity. . . . Such self-deception as this is not reasoning ; it is the negation of reason ; which is the faculty of forming correct conclusions from things observed, judged by the light of experience. It is unworthy of our intelligence and a waste of our

F

greatest gift ; but that introduction serves very well
to illustrate the kind of illusion that lies at the root
of Relativity.

Throughout the whole of his theories there is evidence that Einstein was thinking almost entirely of
their application to astronomy, but it was inevitable
that this should involve him with physics, so that he
had then to engage upon a series of arguments intended to show how his principles would work out on
the plane of general science. The first may be said
to be the motive that inspired him ; while the second
consists of complications and difficulties which he
could not avoid. . . . And when he suggested that
the street might be moving while the waggon with
its wheels revolving was standing still, he was asking
us to imagine that in a similar manner the earth we
stand upon might be moving while the stars that
pass in the night stand still. It is a Case of Appeal,
where Einstein appeals in the name of a convicted
Copernican Astronomy against the judgment of
Michelson - Morley, Nordmeyer, physics, fact, experience, observation and reason. We, on the other
hand, are counsel for the prosecution, judge and jury.

Under the general heading of Relativity, Einstein
includes an assortment of new ideas—each of
which depends upon another,—and each of which
contributes to support the whole. He says
that there is no ether, and that light is a material
thing which comes to us through empty space. Consequently light has weight, and, therefore, is subject to
the law of gravitation, so that the light coming from
a star may bend under its own weight, or deviate from
the straight line by the attraction of the sun, or of
any other celestial body it has to pass in its journey
to the observer on earth. . . . In that case it follows
that no star is in reality where it appears to be, for it
may be even as suggested in diagram 26. . . Consequently the heavenly bodies may be much further
away than they have hitherto been supposed to be,
and every method which is based upon the geometry

of Euclid and the triangulation of Hipparchus will fail to discover the distance to a star ; because its real position is no longer known. Wherefore Einstein has invented a new kind of geometry, in order to calculate the positions of the stars by what is nothing more or less than metaphysics.

We have always been accustomed to measure things by the three dimensions of Euclid—length, breadth and thickness, but Einstein (thinking of astronomy), says that " Time " is a Fourth Dimension ; and proposes that henceforth things should be measured on the understanding that they have four dimensions—length, breadth, time, and thickness.

The introduction of " time " as a fourth proportion of things makes it necessary for him to invent a number of new terms, and also to change the names of some of those that we already know and commonly use, thus, for example—" Space " is changed to "Continuum," while a " point " is called an "event," time—as we have always understood it—no longer exists, and is said to be a fourth dimension ; while there are no such things as " infinity " or " eternity " in relativity.

That is the case for Einstein. It is the essence of his Relativity, clearly stated in plain English. The details of it represent an immense amount of labour of a refined character, the whole thing is very imaginative, and the work of an artist in fine-spun reflections ; indeed, it is of that double-distilled intricacy which finds favour with those who like mental gymnastics and hair-splitting argument ; and are fond of marvellous figures.

But I can conceive that in the course of time this Relative Phantasmagoria might come to be regarded as science, and be taught as such to the children of the near future ; and that is to be prevented only by dealing with it now ! which I will do, though I grieve to give so much space to a matter which only calls for it because it is pernicious.

## Chapter Thirteen

## EINSTEIN'S THEORIES EXAMINED

WHATEVER it is that Relativity is supposed to establish is to be disproved backwards, beginning with the example which Einstein puts forward—where an observer standing at the centre of a rotating disk is watching some one else on the same disk measuring the circumference of a circle round the observer by repeated applications of a small measuring rod; and afterwards measuring the diameter of the circle in the same way.

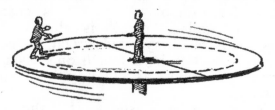

*Diagram* **27.**

He says that because the disk is in motion, the small measuring rod will appear to the observer (at the centre) to be contracted, so that the person who is measuring (whom I will call " B ") will have to apply the rod more often to go round that circle than he would if the disk was at rest. That is not true ! . . . If B actually lays the rod (or foot rule) down upon the disk correctly, the number of applications to go round the circle will be the same whether the disk is moving or not, and the observer at the centre will see that it is so, if he is not made too dizzy to count. On the other hand, if B does not lay the rod down and measure the circle as one would expect, but only walks around the disk with the rod in the air (as in diagram 27) then the rotation of the disk will disturb him, so that he has to make an effort to preserve his balance ; with the result that he can not place the rod as accurately as he would if the disk were not in motion ; and in that case it may take either more or less applications of the rule to go completely round than it would if the disk were still ; and that difference would be seen by the observer at the centre—not as an optical illusion ! (as Einstein implies) but in reality : a result that is entirely physical, and due to physical causes.   When walking across the disk and measuring the diameter, B is not disturbed to anything like the same degree as in walking round the circumference, and so he measures the diameter more accurately. Most of us have at some time or other witnessed the antics of a clown trying to run or walk upon a spinning disk in a circus, and this enables us to understand how such a motion would affect our friends performing on Einstein's revolving table.

His example is merely amusing, it serves no useful purpose, and proves nothing ; unless, indeed, it proves by analogy that the inhabitants on a spinning earth would be rendered as incapable of acting and judging things correctly as his examples.

What we have always known as a " point " in the terms of Euclid, Einstein calls an " event ! " but if words have any meaning a point and an event are two totally different things ; for a point is a mark, a spot or place, and is only concerned in the consideration of material things ; while an event is an occurrence, it is something that happens. . . . There is as much difference between them as there is between the sentence " This is a barrel of apples," and " These apples came from New Zealand."

While claiming " time " as a fourth dimension, Einstein explains that " by dimension we must understand merely one of four independent quantities which locate an event in space." . . . This is to imply that the other three dimensions which are in common use are independent quantities, which is not the case ;

for length, breadth and thickness are essentially found
in combination ; they co-exist in each and every
physical thing, so that they are related—hence they
are not independent quantities. . . . On the contrary, time IS an independent quantity. It is independent of any one, or all, the three proportions of
material things, it is not in any way related ; and
therefore cannot be used as a fourth dimension.

We know that an event is an occurrence ; and we
find that what Einstein really means by his fourth
dimension is " merely the time by which we locate
something that happened in space ; " and that is
just what time has always meant—the period between
one event and another. . . Length, breadth and
thickness, are proportions of each and every finite
thing ; while time is infinite. The dimensions are
finite ; while time is abstract.

.    .    .    .    .    .    .

Strangely enough, while Einstein claims that everything is in motion and nothing is stable, he allows one
thing,  and one thing only,  to remain outside the
realm of relativity, independent of everything else ;
and that is what he calls his Second Law, the
Einstein " Law of the Constancy of the Velocity of
Light." He claims that the velocity of light is constant
under all circumstances, and therefore is absolute.

This is a blunder of the first magnitude, but I do
not imagine that he fell into it through any oversight ;
for it is quite evident that he was driven into this
false position. He was compelled to say that the
velocity of light is constant, because, if he did not
his new geometry would be useless ; for after all his
geometry amounts to this :—

He begins by assuming that light is a material
thing, so that it is affected by the gravitational attraction of any celestial bodies it has to pass on its way to
earth, which causes it to deviate from its appointed
course so that it comes to us with more or less curve,
according to its distance, and according to the

bodies it encounters in its passage. But it always travels at the same velocity, and so, if we can estimate —for example—how much the light of Canopus is made to curve by the gravitation of other bodies between it and the earth (which would be done by Kepler's and Newton's laws), we can calculate how much longer its journey is made by those windings, twists, and turns. Then we can time its arrival, because—although it has to travel so much further than its distance would be in a straight line—it always travels at the same 671,090,400 miles an hour ; or 186,414 miles every second. It is true that Einstein uses a number of signs and symbols which are supposed to simplify the process ; though it is probable that they do no more than merely make it more mysterious, but the plain English of it is as I have shown ; and so we perceive that Einstein uses time pretty much in the same way as we do, and not as a dimension at all.

Thus we have discovered that the things which he re-christened an Event, a Fourth Dimension, and a New Geometry, are false to the titles he has given them ; the words as he uses them are misnomers, therefore we dismiss them ; for they are no longer of any use or interest to us.

Now we are free to deal with his Law of the Constancy of the Velocity of Light.

We are told that Light is a material thing, and that a beam of light is deflected from a straight line by the gravitation of any and every thing that lies near its course as it passes within their sphere of influence ; and we are further assured that light always maintains a uniform speed of 186,414 miles a second. . . . We have, however, to remind Professor Einstein that the " Velocity of Light 186,414 miles a second " was determined as the result of experiments by the physicists—Fizeau, Foucault, Cornu, Michelson and Newcomb, all of which experiments were conducted within the earth's atmosphere, on terra-firma ; the last between Fort Myer and the Washington Monument.

In all these experiments a ray of light was reflected between two mirrors several miles apart, so that it had to pass to and fro always through the atmosphere, and it is not to be supposed that light, or anything else, can travel at the same speed through the air as it would through the vacuum Einstein supposes space to be.

Let us reverse this in order to realize it better. It is not to be supposed that any material thing travels at no greater speed through a vacuum than it does through air, which has a certain amount of density, or opacity. If anything does not distinguish the difference between air and a vacuum, then it is not a material thing; it cannot be matter. On the other hand, anything that is matter must of necessity make such a distinction, and in that case its velocity can not be constant.

Again, if a ray of light can deviate from its course by the gravitational pull of the sun, or of any other celestial body it has to pass, it must accelerate its speed while approaching that body; and slacken it again in reverse ratio after it has passed; hence it follows that its velocity is not constant.

Once more, if a ray of light can bend by its own weight, or by the law of gravitation, it is subject to other conditions, and therefore is not absolute. . . The length of the course used by Newcomb in the final determination of the Velocity of Light was 7.44242 kilometres (return course). If the ray of light had deviated by a hair's-breadth from an absolutely straight line, it never could have passed through the interstices between the very fine teeth of his revolving wheel, or return precisely to the appointed spot on his sending and receiving mirrors, which were 3.72121 kilometres, or more than two and a quarter miles apart in a bee-line. The fact that the ray of light did pass from mirror to mirror, and through the wheel, proves that it maintained a straight line; hence it is certain that it was not deflected from its course by the gravitation of the earth between the two mirrors ; wherefore it is obvious that it was not affected by gravitation.

So we find that the very experiments by which the accepted 186,414 miles per second as the Velocity of Light was measured—experiments which were carried out with the utmost painstaking and minute attention to detail—prove that a ray of light is not influenced by the gravitation of the earth in the slightest degree. Therefore, if those experiments were good enough to warrant all the world in accepting the " Velocity of Light " they may be equally well adduced as proof that a ray of light does not bend by its own weight ; and that light is not affected by gravitation. . . . . And if it is not influenced by gravitation a ray of light cannot be deflected from its course by anything it has to pass, so that its course remains true to the direction in which it was discharged ; and that is a straight line in every direction from the source. (Lord Kelvin tells us that " Light diverges from a luminous centre in all directions.")

In brief—we find that Light is not a material thing, that it is not subject to gravitation, that it has no weight and does not bend, and that it does not describe any kind of curve ; but that it is " an expression," in the same sense as sound is an expression, and that— as such—its velocity varies according to the density of the medium through which it passes ; and that therefore the Velocity of Light is not constant, and Einstein's Second Law is entirely wrong ! . . . . The question of the " ether versus empty space " remains unaffected by his theories, and the stars that glitter like veritable diamonds in the sky are exactly where they appear to be.

. . . . . . .

So much for Einstein's Second Law. Now let us examine the other, the first law, or as he calls it— " The Principle of Relativity"; which states " That all inertial systems, that is, all systems which move with uniform and rectilinear velocity with respect to

each other, are equivalent in expressing the laws of
natural phenomena."

That is what the law is stated to mean. It may not
appear very inviting to the general reader, but he will
find it quite interesting as we proceed, though it is, of
course, of very great importance to every student of
general science and mechanics. As a matter of fact
it is not a law at all, it is a statement. . . At the same
time it is not a plain statement ; for it is equivocal,
and means something which it does not say ; it is a
statement by implication. . . . It is as though we
were to say—" Hello, Jones, how long have you been
out of gaol ? " That would make it necessary for
Jones to prove that he had not been in gaol, in order
to dispose of the implication ; and so it is with this
statement of the Principles of Relativity ; it is an
implication.

Taken literally it is true ; for it states what is
already known ; but it implies the reverse of what it
states—" that all systems which do NOT move with
uniform and rectilinear velocity with respect to each
other are NOT equivalent in expressing the laws of
natural phenomena ! " and that is very much more
important.

Now if we carry this innuendo to its logical conclusion, and put it into simple language, it means—
" that no reliance can be placed upon any deductions
which are obtained by means of observations to the
heavenly bodies, because they are taken from the
surface of the earth, and the observer is moving at a
different speed than the object under observation."

There would be a certain amount of truth in that if
the earth was really moving ; though, even if that
were so, the effects of relative movement could be
easily overcome by taking two observations simultaneously from opposite sides of the meridian to which
the object was vertical. The effects of time would
be eliminated in that way ; and a mean would be
found by comparing the two opposite observations.
And so we find that neither the statement (or law), or

its implication, have any value.   The statement might just as well have never been made.

. . . . . . .

With mental agility worthy of a better cause, Einstein leads from his Mechanical Principle of Relativity up to the Special Principle of Relativity, by means of one of the most extraordinary arguments it is possible to imagine ; but, strange as it is, and inconsequential as it may seem, this argument really affects everything that comes within the range covered by the word " Relativity " ;   and for that reason we will not allow it to pass unnoticed.

After admitting that Electro-magnetic laws do not alter according to the system in which they occur— that is to say—after admitting that Electro-magnetic laws act the same all the world over, he proceeds to argue precisely the contrary, by saying, quite definitely, that in reality they do alter,   and offers to prove it by the following statement :—" The motion of each locality on the earth is constantly changing from hour to hour, but no corresponding changes occur in electro-magnetic action."

Of course this has all the appearance of a man flatly contradicting himself, and it might even appear to be nonsense, but in reality it is a very pretty argument of the most elusive kind which it is a pleasure to meet.   I will confess that I admire Einstein : he skims so close round the edge of the ice. . . .

What he suggests is this :—

The observer is located on the surface of an earth which is rotating on its axis, and at the same time travelling through space at many thousands of miles an hour, consequently his place, or locality, is continually changing with respect to an imaginary point fixed in space.   Notwithstanding this change of place, electro-magnetic laws appear to act precisely as they would if this place was not changing its position with respect to that point.   Therefore Einstein argues that electro-magnetic currents must, in reality, vary their

speed, and so adapt themselves to the changing conditions in such a manner as to "seem the same to
the observer as if he had not changed his position."

Unfortunately he is unable to show any reason why
electro-magnetic action should do this remarkable
thing ; for he treats it as a thing that had intelligence,
as if it wilfully acted in a manner calculated to deceive
the observer.  When reduced to its essence, this argument proves to be no more logical than the idea that
the street might be moving while the waggon was at
rest.  Einstein has been betrayed into supposing a
thing that is altogether impossible , i.e. that a physical
law can act in an unnatural manner, and yet produce an
effect which appears to be normal ;  because he began
by assuming that the locality of the observer was
changing,  and that assumption was untrue !  Now if
he can realize the fact that the earth is actually at
rest, he will find that his difficulties all disappear ;
and that Electro-magnetic laws do not alter, neither
does the locality of the observer change.

But as Einstein persisted in shutting his eyes to the
fact that the earth is stationary he did not see the
incongruity of his assumptions concerning electro-
magnetic action, so that—in order to support his
contention—he was led still further into error, and
compelled to repudiate two of the Laws of Dynamics,
viz. :  1. " Lengths of rigid bodies are unaffected by
motion of the frame of reference ;" and 2, "Measured
times are likewise unaffected."

He says that these two laws of dynamics are untrue,
and thought to prove they were wrong by the fore-
going argument, so it becomes necessary for us to
prove the fallacy of that argument in such a manner
as to leave no doubt whatever as to what is true, and
what is false ; the two " Laws of Dynamics " 1 and 2,
being the stake at issue.

Einstein believes that the earth is rotating on its
axis in the direction of the arrow in diagram 28, at the

rate of 1,000 miles an hour ; and that at the same time
it is travelling, *en masse*, in the same general direction
along its orbit at 66,000 miles an hour ; therefore he
thinks that an electro-magnetic current must travel
from B to A in less time than it will take in travelling
from A to B, because B is all the while running away
from A, while A is always going towards B. . . .
Therefore it appears that the
measured length of a current
passing from B to A (and also
the time it takes) will be shorter
than the measured length and
time of a current passing in the
opposite direction from A to
B; (hence his contention that
lengths of bodies and measured
times must both be affected by
the motion of the observer.)

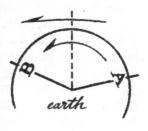

*earth*

*Dia.* **28.**

Of course we know that his premises were wrong,
and that A and B are both located on an earth which
is at rest ; but, for the purpose of the argument, we
will waive that, and assume the Copernican astronomy
to be true. Then his argument is not so unreasonable
as it seemed ; indeed it almost has the appearance of
being true ; but Einstein has forgotten that the
observers at A and B are both on the same earth—
that they both use the same Greenwich Mean Time—
and that the Electro-magnetic wave passes from one
place to the other by convexion—so that the earth's
atmosphere offers the same facility to its passage
from A to B, as it does from B to A.

And that is the trifle that turns the scale against
him. The fact that the whole operation takes place
within the terrestrial atmosphere gives equal conditions to an electro-magnetic current passing in any
direction within that atmosphere ; the same being
unaffected by anything that may, or may not, take
place in ethereal space, which the earth and its
atmosphere in its entirety is unconscious of . . . .
Thus, an electro-magnetic wave passes from A to B

in the same time as it passes from B to A, just as a train travelling at a uniform speed of 60 miles an hour goes from Bristol to London in the same time as it will go from London to Bristol ; while the length of the railway track measures the same from Bristol to London as it does from London to Bristol.

And so the Laws of Dynamics 1 and 2 remain true ; while Einstein's contention has been proven false.

The whole hypothesis of Relativity has failed, both in the mass and in detail, under our examination, so that, unable to support itself, it can no longer aspire to support any theory of the universe. Therefore our judgment remains unaltered. Copernican Astronomy stands condemned, and has lost its last, and perhaps its ablest, living advocate.

## Chapter Fourteen

## EINSTEIN'S EVIDENCE

BUT it will be remembered that he offered three crucial tests as evidence in support of his theories, and these we have still to examine. They are :—

1. That certain irregularities in the movements of the planet Mercury would be accounted for by Einstein's geometry.

2. That because light has weight it would bend by gravitation as it passed near another body on its way to the earth, and that this could be verified by observations taken at the time of a solar eclipse.

3. That certain lines in the spectrum would be found to shift.

We have done with mental athletics, and here we have something a little more tangible to deal with.

Of the Third it is said by the Authorities of Astronomy that the observations necessary to prove or disprove such a shifting of the lines in the spectrum would be so extremely difficult that it is practically impossible ever to do it, and therefore it is set aside.

.    .    .    .    .    .

The First is very well handled in an article by T. F. Gaynor in the London *Daily Express* of June 6th, 1921.

Mr. Gaynor meets Einstein on his own ground as a good astronomer should, and uses figures which take my breath away ; but, nevertheless, I will leave him to deal with crucial test number 1.

He says that the discovery of Neptune, 75 years ago, by means of Newton's Law, utterly extingiushes the Einstein theory so far as Mercury is concerned.

Irregularities similar to those of Mercury had been observed in the movements of Uranus, and in 1841 it was thought that these unaccountable movements must be due to the gravitation of some other planet at that time still undiscovered. But I will quote Mr. Gaynor verbatim :—" Uranus is 1,800,000,000,000 miles from the sun. Adams and Leverrier, applying Newton's Law, which, according to Einstein is an exploded theory, located the probable position of the undiscovered planet a thousand million miles still further on in space—and there Dr. Galle, the Berlin astronomer, found it, on September 23rd, 1846.

Thus, 75 years ago, the Newtonian law found a previously unknown planet (Neptune) at a distance of 2,800 millions of miles from the sun, yet Einstein would have us believe that the same law does not hold good with regard to Mercury ; which is only 36,000,000 miles from the sun ! . . . The " proof " he adduces from the aberration of the orbit of Mercury can be disposed of in a sentence. He has made the elementary blunder of regarding Mercury as globular instead of spheroidal."

## LIGHT AND GRAVITY.

There remains now but one last defence of the Theory of Relativity, and that is the statement that light is really matter, and that it is subject to gravitation. (Test No. 2.)

In order to put this to the test, expeditions of British Astronomers were sent to Sobral in North Brazil, and to the island of Principe on the west coast of Africa, to observe the total eclipse of the sun on May 29th, 1919, and the results they obtained seemed to justify Einstein's main test, so that as a consequence the Royal Astronomical Society held a remarkable meeting at Burlington House on November 6th, 1919 ; and on the next day all the world of astronomy did homage to Einstein.

The results of the eclipse appeared to satisfy the gathering at Burlington House. Sir Frank Dyson, the Astronomer Royal, described the work of the expeditions, and convinced the meeting that the results were definite and conclusive. Dr. Crommelin explained that the purpose of the expeditions was to test whether the light of the stars that are nearly in a line with the sun is bent by its attraction, and if so, whether the amount of bending is that indicated by the Newtonian law of gravitation, viz. : seven-eighths of a second at the sun's limb, or the amount indicated by the new Einstein Theory; which postulates a bending just twice as great. . . . The results of the observations were 2.08 and 1.94 seconds respectively. The combined result was 1.98 seconds, with a probable error of about 6 per cent. This was a strong confirmation of Einstein's Theory, which gave a shift of 1.75 seconds.

The fourth dimension was discussed, and it appeared that Euclidian straight lines could not exist in Einstein's space. All lines were curved, and if they travelled far enough they would regain the starting point. Mr. de Sitter had attempted to find the radius of space. He gave reasons for putting it at about a billion times the distance from the earth to the sun, or about sixteen million light-years ! This was eighty times the distance assigned by Dr. Shapley to the most distant stellar cluster known. The Fourth Dimension had been the subject of vague speculation for a long time, but they seemed at last to have been brought face to face with it.

Even the President of the Royal Society, in stating that they had just listened to " one of the most momentous, if not the most momentous, pronouncements of human thought," confessed that no one had yet succeeded in stating in clear language what the theory of Einstein really was. . . . But he was confident that " the Einstein Theory must now be reckoned with, and that our conceptions of the fabric of the universe must be fundamentally altered."

G

Subsequent speakers joined in congratulating the observers, and agreed in accepting their results. More than one, however, including Professor Newell, of Cambridge, hesitated as to the full extent of the inferences that had been drawn, and suggested that the phenomena might be due to an unknown solar atmosphere further in its extent than had been supposed, and with unknown properties.

With such a reception as this it is not surprising that the followers of Copernicus everywhere should be almost willing to believe in Relativity whether they understood it or not ; but the Royal Astronomical Society might have been a great deal more careful than they were, as we shall see :—

That the Einstein Theories were automatically coming to be regarded as accepted science, is evidenced by the fact that the Astronomer Royal himself introduced them into a public lecture on eclipses which he gave at the Old Vic. in the February of 1921.

Coming to the description of the eclipse of May 29th, a slide was thrown upon the screen to illustrate the result of the observations that were said to verify Einstein's ' Theory. (See diagram 29.)

The lecturer described how certain stars which were in the same direction as the sun could, of course, not be seen in the ordinary way in the

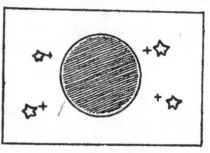

Diagram 29.

day time, but when the sun was obscured, as at the time of a total eclipse, they could be seen through a smoked glass or telescope. The exact position of these stars was known to astronomy, but if Einstein's Theory was correct the light coming from them to the observer would be bent as it passed near the sun, so that they would not appear to be in their true positions. Then

he showed how the Einstein Theory was verified ;
for the stars were observed to be a little further from
the sun than their theoretical or true positions.

But the Law of Gravitation is " That mutual action
between masses of matter by virtue of which every
such mass tends towaid every other, &c., &c."

Observe that it tends toward ;   it attracts ;   it
pulls :  therefore—if light was matter, and was affected
by the gravitation of the sun, the stars would be seen
nearer to the sun ;  and not as stated by the lecturer
and illustrated on the slide.

In diagram 29 the crosses XX suggest the normal,
true, or theoretical positions of the stars with respect
to the sun.   If Einstein's theories had been right the
stars would be seen nearer to the sun than the crosses,
but the Astronomer Royal demonstrated the fact
that they were actually further away !

Such was the real result of the solar eclipse of May
29th, 1919.   The circumstances had been laid before
the Royal Astronomical Society in Burlington House
on November 6th, and yet, for some unaccountable
reason they failed to perceive that the result was
contrary to the Law of Gravitation ;  and clearly
demonstrated the fact that Einstein's Theory is false.

. . . . . . .

N.B.—The real cause of the displacement of these
stars from their true positions is known to
the author, and will be explained in a
book sequel to this work ;  but he does not
consider that explanation necessary to the
present discussion.   Einstein's Theory is
disproved ;  alternative or no alternative.

## Chapter Fifteen

# MARVELS OF ASTRONOMY

NOTHING now remains of that astronomy which was once said to be the most perfect of the sciences ; and imagination—stretched even to its uttermost—has failed to support it in the face of reason, and yet these last two years since Relativity became the vogue have produced the most remarkable figures astronomy has ever known.

## " BETELGEUSE."

In December 1920, Professor Michelson related how he had perfected an instrument known as an Interference-Refractometer, and how he had used it to measure the angular diameter of the star Betelgeuse, in the Belt of Orion ; and found it to be 0.046 seconds of arc. That is to say that he found the measurement of this star as it appears to the eye (which is only like a glittering pin-point) to be 0.046″ from one side to the other, and that is one-twentieth part of a second of arc, or 1-72,000th part of a degree ; very fine measurement indeed.

Professor Michelson, however, is a physicist, specially interested with theories of light, and so, having invented the instrument and measured the apparent diameter of the star, his work was done.

Astronomers then took up the matter, and on referring to their records, found the distance of Betelgeuse to be 180 light-years ; that is 180 times 6,000,000,000,000 miles, or one thousand and eighty billions of miles from the earth ; and so they calculated that if a thing so far away appeared to be 1-72,000th part of a degree in diameter, its real diameter must be two hundred and sixty million miles !

Then the world of astronomy pointed with pride to the mighty star that was 260 million miles from one side to the other, and told how the sun was a million times bigger than the earth, while Betelgeuse was 27 million times bigger than the sun. . .

The actual size of Betelgeuse, however, depends upon its distance, and as we have shown in the chapter on " 61 Cygni " that the astronomers' method of measuring stellar distance is absolutely useless, we know that they are entirely wrong in supposing Betelgeuse to be 1,080 billions—or any other number of billions—of miles from the earth. Therefore it follows that as they do not know its distance, they may not use its apparent diameter and divide that into unknown billions of miles. Being in reality quite ignorant of the distance of Betelgeuse, they have no legitimate means of forming any conception of its dimensions at all. Those dimensions are to be ascertained by first finding the star's real distance, which is something less than twenty thousand miles. Then that may be divided by Professor Michelson's 0.046", which will show the actual size of that twinkling little point of light known as " Betelgeuse " to be not much more than twenty-five feet !

.     .     .     .     .     .     .

It has since transpired that the distance to Betelgeuse had been measured on three different occasions, each time with a different result. One of these showed it to be 654 billions, another made it 900 billions, while the other gave it as 180 light-years, or 1,080 billions of miles away ; and it is surprising that astronomers did not realise the fact which was clearly demonstrated by these differences—that their methods of measuring stellar distance are not to be relied upon.

In the meantime we can see no reason why they

preferred to use the greatest of the three various
estimates of the star's distance—in conjunction with
Michelson's angular diameter—rather than the least,
for that only seems to have had the effect of magnify-
ing the dimensions of Betelgeuse to the uttermost.

## "PONS-WINNECKE."

While the excitement over Betelgeuse was at its
height the universe loomed even larger than before,
for Canopus and Rigel were then said to be " 460
light-years away and they may be 1,000 or more."
Meanwhile Dr. Crommelin gave us a scare with the
story of how a comet called Pons-Winnecke was
rushing toward the earth at a hundred thousand miles
an hour, while Dr. Slipher dicovered a nebulous mass
that was gyrating round the firmament at eleven
hundred miles a second ! ! ! This, so far, has never
been surpassed, and " SPIRAL NEBULA NUMBER
584 " still holds the record of being the fastest thing
in creation ; its velocity being so great that it could
go from Liverpool to New York in two ticks of the
clock.

Pons-Winnecke had been seen somewhere in Africa
in January 1921, and it was predicted that this comet
would be visible at London in June ; and this gave
rise to much speculation. It was said that Pons-
Winnecke might strike the earth with a fearful bump
about the 26th of June, but Mr. E. W. Maunder said
that though there might be a bump it is only a fog of
gas after all ; while Dr. Crommelin thought the comet
might miss the earth this time, and so there appeared
to be no danger. . . Then Sir Richard Gregory said
that if the head of Pons-Winnecke did hit the earth
it might set the world on fire, but we were reassured
again when he told us that there is about as much
chance of the comet hitting the earth as of a random
shot hitting a bird in full flight ; yet it seemed strange
that he should imagine a comet to be like a random

shot in this well-ordered universe ; unless, perchance, he had forgotten about the Law of Gravitation.  And how are we to understand how the earth could be set on fire when he tells us that we may pass through the tail of a comet without harm because it is really a far higher vacuum than anything that can be produced in our laboratories ? . .  Then what are we to think of it all when Professor Fowler tells us that we don't know how a comet is formed, we don't know where it comes from, and don't seem really to know what it is ? . .  He thought they may come from gases thrown off from the sun which are gradually cooled ; but that made it even more difficult to understand how it could set the earth on fire, or what all the bother was about.

Nevertheless the discussion continued, until at last the leading authorities advanced the " Fascinating Theory that Pons-Winnecke may have come from a distance in space so great that it is impossible to think or speak of that distance in terms of miles."  That took our breath away, for it appeared that the comet might come out of illimitable space, to wander amid the stars at its own sweet will, regardless of the Laws of Dynamics and Gravitation. . .

Even yet the romance is not complete—for after waiting in great expectation for several months the Secretary of the Royal Astronomical Society told us that " Pons " had been seen again ! this time with only a stump of his original tail, though even this stump was five hundred million miles long, and seemed to be comprised mostly of gas and meteors. . .  It is not recorded how he knew the length of its tail, and nothing was said as to what had become of the remainder ; but to cut a long tale short—the summer came and passed—but Pons-Winnecke never arrived ! . . . He was lost ; and even now he may be wandering on and on, somewhere in fathomless space, no one knows whither ; and nobody cares.

## "THE RUDDY PLANET."

At about the same period there was much ado about the planet MARS.

It had long been supposed that this planet was very much like the earth, but inhabited by a race of giants, probably about fifteen feet in height. Some straight lines which had been observed on the planet were thought to be irrigation canals made by men ; and one could imagine fields of cabbages, cauliflowers, and spring onions growing along the banks ; indeed one could imagine anything. And so, when wireless operators in various parts of the world began to hear strange noises which they could not account for (about the time of Pons-Winnecke) the rumour spread abroad that they might be wireless signals from Mars.

It was not suggested that the Martians might be sending these signals in reply to those we had thought of flashing to them in 1910, but it was supposed that the people on Mars might have been hearing things ; and thought our wireless operators were tic-tacking to them. So the possibility of sending messages to the ruddy planet by wireless telegraphy came to be discussed almost as much as the comet.

Astronomers said that although the earth is about seventeen million years old, Mars is very much older ; therefore it was presumed that the Martians would probably be more advanced in knowledge than we are, and might have been using wireless for goodness knows how long, and had now discovered that we had a Marconi System.

The tappings and cracklings that were heard sometimes at night were rather uncanny, and could not be understood, but this was not because the Martian's language was different than ours ; it was because the vibrations that affected the wireless coherers were really caused by the splitting of the ice around the pole !

Spring was advancing in the northern hemisphere, and the ice-fields were melting and breaking before

the warmth of the advancing sun, so that the colliding
and shifting of huge bergs disturbed the normal
distribution of the magnetic currents from the north
Pole. . . .

Professor Pickering might have made this discovery
if he had had time to think of it ; but at that period
he was busy studying the weather of Mars. I don't
think he knows any more about the weather on earth
than the Meteorological Office, but I recollect that he
told us it was snowing on that little old planet ; and
that was a very remarkable thing, if it was true—
indeed it was remarkable whether it was true or not.
Time was when it was said that water ran uphill
instead of down on Mars, and in the year A.D. 1910,
all sorts of schemes were proposed for signalling to
the planet by means of bonfires and search-lights at
night, or by using mirrors to reflect the sun's rays by
day. It was all very interesting in its way, but very
nonsensical—because the sun is always shining on
that side of Mars
which is presented to
us, whether it is day
or night on our side
of the earth ; and so
it would be impos-
sible for the Martians
—if there were any—
to see our bonfires or
our mirrors, because
with them it must
always be daylight,
and they could not
even see the earth
itself ! . . .

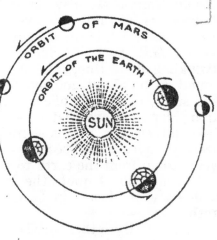

Diagram 30.

This is because Mars
goes round the sun on
a greater orbit than
the earth, while we travel on the inner circle, according to the Heliocentric Theory, (as shown in diagram 30).

It is surprising that astronomers had not thought of this, but they will find that it is so, if they will only study their own astronomy.

But the time has come when all the romantic things that have been said about Mars must take their proper place among fairy tales, for if the distance to that planet is measured by two simultaneous observations, as I have advised for the measurement of the sun, it will be found to be never more than 15,000 miles from the observer, and too small altogether to be inhabited ; too small even for Robinson Crusoe and his man Friday. . . .

## " N.G.C. 7006."

Before bringing this history of the evolution of modern astronomy to a close I have yet to mention the constellation of Hercules, which Dr. Shapley at Mount Vernon recently estimated to be about 36,000 light-years distant, or 200 times further off than Betelgeuse ; while we are now told that a star known as " N.G.C. 7006 " (which is one of those myriad twinkling little things in the Milky Way) has been found to be about 200,000 light-years distant ; and this surely is the limit of even an astronomer's imagination ; for it means that it is so far off that it would take an electric current—travelling at the rate of 186,000 miles every second—two hundred thousand years to go from the earth to the Milky Way ! . . .

In conclusion I quote the following from an article which was published in London as recently as April 15th, 1922 :—

" . . . . By other methods most bodies in the heavens have been measured, and even weighed, and the results obtained stagger imagination. One of such methods consists in watching an object through the spectroscope and making calculations from the shifting of the lines in the spectrum. In this way the mighty flames which leap from the surface of the sun have been measured. Some years ago one flame was observed to shoot out with a velocity of at least 50 miles a second, and to attain a height of 350,000 miles ! . . . The stars in general cannot be measured ; but the thing has been done in some cases, notably by Bessel, who, after three years' observations of 61 Cygni, announced its approximate distance from the earth as not more than sixty billion miles ! Yet this is one of our nearest neighbours among the distant suns. It is so close to us—comparatively—that we have learned a lot about it since Bessel made his calculations.

Scientists have shown that a difference of a mere twenty billion miles in distance from the earth is negligible, and that, though it is tearing through space at thirty miles a second, it would require about forty-thousand years to make a journey equal to its distance from the sun."

It is difficult to tell whether the journal was joking or not ; it appears to be so, but, nevertheless, the statements are those given out in all seriousness in the name of Astronomy. They are the things which are being taught in colleges and schools as scientific knowledge in this month of May, 1922 ; for which astronomers, the Educational Authorities, and the indifference of parents are responsible.

However, it is to be observed that—with the single exception of Alpha-Centauri—since Bessel estimated the distance of the first star to be sixty-three billion miles away, stellar distances have grown greater and greater, until at last we have this " N.G.C. 7006," said to be twenty thousand times further than 61 Cygni ! or " one million two hundred thousand billions " of miles from this earth of ours.

And this preposterous figure is the outward and visible sign of the nature of the science that has been evolved in twenty centuries through the failure of astronomers to perceive the error of Hipparchus.

Adieu.